Injection
Electroluminescent Devices

Injection
Electroluminescent Devices

C. H. GOOCH

Services Electronics Research Laboratory
Baldock, Hertfordshire

A Wiley–Interscience Publication

JOHN WILEY & SONS
London · New York · Sydney · Toronto

Library of Congress Catalog card No. 72-5716

ISBN 0 471 31280 0

Printed in Great Britain by
Dawson & Goodall Ltd.,
The Mendip Press, Bath

Preface

Electroluminescent diodes and lasers are semiconductor p–n junction devices that emit optical radiation when operated under forward-bias conditions. Although the term 'electroluminescent' strictly implies the emission of visible radiation, it is usually extended to include emission in the near-infrared region of the spectrum, since the technology involved is similar in both cases.

At present, most of the devices that are obtainable emit in the red or infrared, while green-emitting devices are also becoming more readily available. Devices operating at shorter wavelengths, in the blue or ultra-violet, have not yet shown a useful efficiency, but research aimed in this direction is continuing in parallel with that aimed at improving the devices now available.

Ten years ago, electroluminescent diodes and semiconductor lasers were still in the experimental stage; now they are practical devices with many applications. The aim of this book is to provide a basic understanding of these devices and their applications. The theoretical and experimental work that is considered is restricted to that which has a direct bearing on practical devices, and no attempt has been made to give a comprehensive account of all the work that has been reported. It is thus hoped that this book will supply a need which is likely to grow as the use of electroluminescent devices becomes more widespread.

I would like to thank my colleagues, in various laboratories, for reading the manuscript of this book, and making valuable comments. I am also grateful to my wife, Elizabeth, for her help in the preparation of the manuscript.

Baldock, Hertfordshire   C. H. GOOCH

v

Contents

1

Introduction

1.1 Electroluminescent diodes and semiconductor lasers

An electroluminescent diode is a semiconductor device that emits optical radiation when an electric current passes through it. In its narrowest sense, the name implies that the radiation emitted falls in the visible region of the spectrum, and devices that meet this criterion are often known as light-emitting diodes or l.e.d.s.

It is often convenient to broaden the term to include not only devices that emit visible radiation, but also those that emit in the near-infrared region of the spectrum. In general, the principles and technologies of these devices are very similar, the main difference being in the applications to which the devices are put.

It might also seem logical to consider devices that emit in the ultraviolet region of the spectrum. However, this is not necessary in practice, since the range of available devices extends, on the short-wavelength side, only to green-emitting devices. There is, however, a considerable research activity aimed at developing devices that operate at shorter wavelengths.

As a further extension, it is appropriate to consider devices that emit stimulated radiation, i.e. the semiconductor p–n junction laser, often referred to as the injection laser. These devices emit in the visible and near-infrared spectral regions, although the most developed and widely available semi-conductor lasers are based on gallium arsenide and emit in the spectral region near 0·9 μm.

All these devices are based on semiconductor p–n junctions formed in single-crystal semiconductors. This is in contrast to other devices, which consist of thin layers of the semiconductor zinc sulphide deposited on suitable electrodes. These devices emit visible radiation under suitable a.c. or d.c. excitation, and have found a number of applications. However, the technology of these devices is very different from that of the single-crystal diodes, and the way in which they operate is, as yet, little understood. For these reasons, it is appropriate to restrict the subject matter of this book to p–n junction devices. This is not to dismiss the usefulness of the ZnS panel devices, but to recognize that a consideration of these requires a different and more empirical approach that is outside the scope of this book.

1

1.2   Historical introduction

1.2.1   Early work

The earliest observations of electroluminescence were made as a result of experiments on detectors for radio waves. As early as 1907, Round reported that yellow light was produced when a current was passed through his silicon carbide detector.[1] Similar results were observed by Lossev in 1923, again when working on carborundum (i.e. SiC) detectors.[2] Lossev made a systematic study of this effect and published his results in a series of papers over a period of years up to 1940.[3] Lossev clearly recognized that the effect he observed was connected with a rectification process.

The phenomenon observed by Lossev is the basis of the work described in this book, and is still often known as the Lossev effect.

The Lossev effect should be distinguished from the Destriau effect.[4] Destriau's work involved zinc sulphide phosphors, and he observed that these phosphors could emit light when excited by an electric field. He termed this effect 'electroluminescence' and the phenomenon that he observed is the basis of the a.c.-excited phosphor panels that have been developed.

The Lossev and Destriau effects are thus the basis of the two classes of electroluminescent devices mentioned in the previous section.

Our understanding of the mechanism of the Lossev effect stems from the work of Lehovec, Accardo and Jamgochian, who proposed a model based on the band-structure diagram of a p–n junction.[5] In their model, electrons are injected across a forward-biased p–n junction and combine with holes in the p-type region of the junction. In this recombination process, the energy lost by the electron is emitted as a photon. Although many refinements have, of course, been added to this simple model, it still forms the basis of our understanding of the electroluminescent effect in p–n junctions.

1.2.2   Electroluminescent diodes

The successful development of electroluminescent diodes can, to a large extent, be related to the successful study and exploitation of the III–V group of semiconducting compounds, although other materials have also played a part and may become of greater importance in the future.

The development of electroluminescent devices cannot be divorced from the development of other semiconductor devices. In particular, the development of the transistor has had a marked influence, partly in providing an impetus for a wide range of basic semiconductor work, and also in making available a highly sophisticated technology which has been applied to electroluminescent devices.

The first transistors were made from germanium. These had severe limitations in the temperature range over which they could operate and have now been almost entirely replaced by silicon transistors. Silicon has a wider band gap than germanium, so that silicon devices can operate over a wider temperature range; but the material suffers from the disadvantage that it has a lower electron mobility (Table 1.1) than germanium.

TABLE 1.1. Some properties of germanium, silicon, gallium arsenide and gallium phosphide at room temperature.

Property		Germanium	Silicon	Gallium arsenide	Gallium phosphide
Band gap (eV)		0·8	1·1	1·4	2·2
Carrier mobility	Electron	3900	1500	9000	200
(cm²/V s)	Hole	1900	600	500	150

In the early 1950s, Welker pointed out that the III–V compounds had semiconducting properties, and considerable effort has since been devoted to a study of the materials.[6] In particular, gallium arsenide attracted a great deal of attention as its basic properties (band gap and electron mobility) suggested that it would be a transistor material that would be superior to silicon. However, gallium arsenide has not replaced silicon as a transistor material, and there now seems to be no prospect that it ever will. This is largely due to the great advances that have been made in the technology of silicon devices, such as the development of planar techniques and the ensuing microelectronics technology. On the other hand, while silicon was realizing its great potential in this field, progress with gallium arsenide was hampered by the difficulties encountered in preparing gallium arsenide of a purity adequate for transistors.

There is also a fundamental difference between silicon and gallium arsenide which is relevant to their use both in transistors and in a wider range of semiconductor devices. This difference concerns the nature of the band structure of the materials, and will be considered in more detail in subsequent chapters. The full implications of this difference were not realized in the early 1950s, but it is now clearly recognized that gallium arsenide (which is a direct band-gap material) has important properties not possessed by silicon or germanium (which are indirect band-gap materials), and it is these that are largely responsible for the continued interest in gallium arsenide.

One of the ways in which this difference is manifested is in the nature of the recombination processes which occur in a semiconductor. For example, when an electron is injected into a *p*-type semiconductor, it will recombine with a hole. In a direct band-gap semiconductor there is a high probability that the energy released in this process will give rise to a photon (radiative recombination), whereas, in the indirect band-gap materials, this probability is much lower, and non-radiative recombination processes will usually dominate.

During the 1950s, the research on gallium arsenide gained momentum, and it was during this period (in 1955) that Braunstein[7] made the first observations of radiative recombination in the III–V compounds. However, it was not until 1962 that Pankove[8] showed that gallium arsenide diodes could be efficient sources of infrared radiation.

Meanwhile work had continued on the other III–V compounds, and a number of workers had observed radiative recombination in the visible region of the spectrum in gallium phosphide.[9-12] The efficiency of these diodes was low (typically less than 0·1 per cent.), as could be expected, since gallium phosphide has an indirect band gap. Nevertheless, their efficiency was sufficiently high for the devices to find a number of viable applications.

During the 1960s, much of the work on electroluminescent devices was aimed at obtaining efficient devices operating in the visible region of the spectrum. For example, it was hoped that the high efficiencies shown by gallium arsenide devices in the near infrared might be also obtained with devices that emit in the visible region of the spectrum. This hope has by no means been fulfilled, but considerable progress in this direction has been made.

The research aimed at efficient visible electroluminescence can be considered as two rival research programmes.

The most obvious approach to the problem is to find materials that retain the direct band-gap structure of gallium arsenide, yet have a band gap that is sufficiently large to give rise to visible electroluminescence. With this approach, the work on the alloy material gallium arsenide–phosphide has been particularly successful in giving devices that emit in the red region of the spectrum, and there are prospects that other similar materials will extend the spectral region available.

The other approach has been to pay careful attention to the purity and other properties of the indirect band-gap material gallium phosphide. By careful control of the elements used to dope gallium phosphide and the careful exclusion of unwanted impurities, it has been shown to be possible to overcome, to a large extent, the disadvantages due to the indirect band gap of the material. In this way a range of useful electroluminescent devices has been made in this material.

The situation at the present time is that $Ga_xAs_{1-x}P$ and GaP devices are competing strongly to meet the requirements for electroluminescent devices. The relative merits of these two approaches are considered in Chapter 4.

1.2.3   Semiconductor lasers

The first semiconductor lasers were made in 1962 by three research groups working in the United States.[13-15] At this time, the principle of the laser was

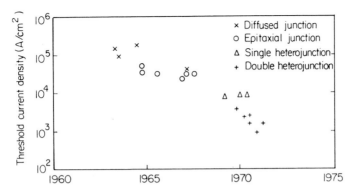

FIGURE 1.1. The development of gallium arsenide lasers, showing
the progressive reduction of reported threshold current densities
at room temperature since 1963

already well established in ruby and some gases. Furthermore, work on
gallium arsenide had shown that gallium arsenide diodes were efficient
sources of radiation. Thus, with hindsight, the development of a semiconductor laser seems to have been a natural extension of available knowledge
and technology.

The first semiconductor lasers were made of gallium arsenide, and much
of the subsequent work has utilized this material. These early devices had
very high threshold current densities and, for practical purposes, could only
be operated at liquid-nitrogen temperatures or below. At this time, the
p–n junctions used were invariably made by a diffusion technology, and
natural developments of this made considerable improvements to lasers.
However, more dramatic improvements, which led to practical devices
operating at room temperature, have depended on the use of increasingly
sophisticated epitaxial techniques to form p–n junction structures. These
are considered in detail in Chapter 5, but the progress which has been made
is illustrated in Figure 1.1. This figure shows, schematically, the way in
which the threshold current of devices reported in the literature has decreased
over a period of ten years. The evolution of a particular technology and the
more dramatic improvements produced by the introduction of new technologies can be clearly seen. These have resulted in a reduction in the room-
temperature threshold current by two orders of magnitude, and one is
tempted to ask where the next innovation will occur.

1.3 References

1. T. Round, *Electrical World*, 309 (1907).
2. O. V. Lossev, *Telegraphia i Telefonia*, **18**, 61 (1923).
3. O. V. Lossev, *Phil. Mag.*, **6**, 1024 (1928).

4. G. Destriau, *J. Chim. Phys.*, **33**, 587 (1936).
5. K. Lehovec, C. A. Accardo and E. Jamgochian, *Phys. Rev.*, **83**, 603 (1951).
6. H. Welker, *Z. Naturforsch.*, **7a**, 744 (1952).
7. R. Braunstein, *Phys. Rev.*, **99**, 1892 (1955).
8. J. I. Pankove, *Phys. Rev. Lett.*, **9**, 283 (1962).
9. G. A. Wolf, R. A. Herbert and I. D. Broder, *Phys. Rev.*, **100**, 1144 (1955).
10. H. G. Grimmeiss and H. Koelmans, *Phys. Rev.*, **123**, 1939 (1961).
11. M. Gershenzon and R. M. Mikuyak, *J. Appl. Phys.*, **32**, 1338 (1961).
12. J. Starkiewicz and J. W. Allen, *J. Phys. Chem. Solids*, **23**, 881 (1962).
13. R. N. Hall, G. E. Fenner, T. J. Soltys and R. O. Carlson, *Phys. Rev. Lett.*, **9**, 336 (1962).
14. M. I. Nathan, W. P. Dumke, G. Burns, F. H. Dill and G. J. Lasher, *Appl. Phys. Lett.*, **1**, 62 (1962).
15. T. M. Quist, R. H. Rediker, R. J. Keyes, W. E. Kragg, A. L. McWhorter and H. J. Zeiger, *Appl. Phys. Lett.*, **1**, 91 (1962).

2

The Theory of Electroluminescent Diodes and Lasers

2.1 Introduction

The theory of electroluminescent devices that is given in this chapter is not an exhaustive or rigorous treatment of the subject, but is intended to form a basis for an understanding of the way in which practical electro-luminescent devices operate. Furthermore, the treatment presupposes a knowledge of the fundamentals of semiconductor physics such as the application of Fermi–Dirac statistics, the concept of holes and electrons and the elementary theory of p–n junctions.

Electromagnetic radiation may be produced when a physical system relaxes from a state of high energy to one of lower energy. If the energy difference between these states ΔE is all converted to a single quantum of radiation, the wavelength of this radiation is given by the familiar expression

$$\Delta E = \frac{hc}{\lambda} \tag{2.1}$$

where h is Planck's constant and c is the velocity of light. Such transitions occur, for example, in an excited gas. In a gas, the energy difference between the two states is closely defined, and the radiation that is produced is emitted as a line spectrum. It can be seen that there are two basic requirements which must be met in a system which is to produce radiation.

The first requirement is for a physical system which can be excited and will then relax to a lower-energy or ground state with the emission of a photon of the required wavelength. It may, of course, be possible for an excited system to lose energy other than by the emission of the required photon. The requirement of the system then becomes much more stringent in that the probability of the required radiative transition occurring must be greater than that of any non-radiative transition which can occur. This point will be considered in more detail in a subsequent section.

The second requirement is for a means of exciting the system. In general, this can be achieved in a number of ways, of which the simplest is, perhaps, an increase in temperature. However, in an electroluminescent diode the system is excited directly by the passage of an electric current through a semiconductor p–n junction. Heating does not play an essential part in this excitation process, although it may exist as an unwanted secondary effect.

In the following sections it will be shown how these requirements can be met in semiconductor devices.

2.2 Radiative and non-radiative transitions in semiconductors

A representation of the band structure of a typical semiconductor is shown in Figure 2.1. It will be convenient to consider p-type material, but an equivalent discussion can be applied to n-type material.

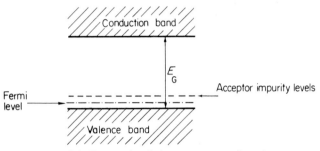

FIGURE 2.1. Energy levels in a p-type semiconductor

The valence and conduction bands are separated by the band-gap energy E_G and, in thermal equilibrium, there will be a high density of holes in the valence band, but extremely few electrons in the conduction band. In p-type material the holes are known as majority carriers and the electrons are known as minority carriers. If now, by some means, an excess electron is injected into the system, occupying an energy level in the conduction band, the thermal equilibrium will be disturbed. Equilibrium will be restored if the injected electron combines with a hole, and the energy released when this occurs can appear as a photon.

It is frequently convenient to consider the rate at which such recombination processes occur in a semiconductor. Considering transitions between the conduction and valence bands in a semiconductor that is in thermal equilibrium, the principle of detailed balance shows that the rate at which electrons are thermally excited from the valence band R_{vc} must equal the rate at which electrons return to the valence band R_{cv}. If n_0 and p_0 are the equilibrium concentrations of electrons and holes

$$R_{cv} = Bn_0p_0 = R_{vc} \qquad (2.2)$$

where B is a constant.

If equilibrium is disturbed by the injection of electrons at a rate g, the steady-state excess density of electrons produced can be shown to be given by

$$\Delta n = \frac{g}{B(n_0 + p_0)} = g\tau \qquad (2.3)$$

where τ is defined as the lifetime of the excess carriers.

If the generation ceases, the excess concentration of electrons will decay as

$$\Delta n = g\tau \exp(-t/\tau). \tag{2.4}$$

For example, the value of the constant B in gallium arsenide has been estimated[1] as 10^{-9} cm^3/s; so that the lifetime of an excess electron injected into p-type material with a hole concentration of 10^{18} cm^{-3} is approximately 10^{-9} s. By comparison, the minority-carrier lifetime in semiconductors such as silicon is several orders of magnitude longer than this.

A similar analysis can be applied to more complex energy-level systems, but a discussion of such systems is beyond the scope of this book, and the reader is referred to the standard texts on semiconductor physics.

2.2.1 Radiative recombination processes

The recombination process can occur with the emission of radiation in a number of basic ways, as illustrated in Figure 2.2.

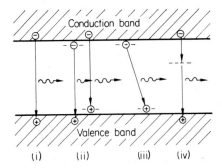

FIGURE 2.2. Radiative transitions: (i) band–band, (ii) via shallow donor or acceptor levels, (iii) donor–acceptor and (iv) via deep levels

[i] *Band-to-band recombination.* If an electron in the conduction band recombines directly with a hole in the valence band, a photon of energy equal to or greater than the band-gap energy E_G of the semiconductor will be produced. The electron and the hole will recombine from states close to the band edges, but the thermal distribution of carriers among these states will give an appreciable width to the emission spectrum. In general, band-to-band recombination will only be observed in pure materials.

[ii] *Recombination via shallow donor or acceptor levels.* In this case, the energy of the transition can be somewhat smaller than the band-gap energy. However, donor and acceptor levels can be very close to conduction and valence bands, often being separated from them by only a few millielectron-volts. Thus it may only be possible to differentiate between shallow impurity transitions and band-to-band transitions by sophisticated spectral measure-

ments at low temperatures. Nevertheless, the participation of an impurity level in a transition can affect the probability of the transition by allowing an effective relaxation of momentum selection rules. This point will be considered in more detail in the next section.

[iii] *Donor–acceptor recombination.*[2] An electron trapped on a donor can recombine with a hole trapped on an acceptor that is several lattice spacings away. The energy involved in this transition will depend on the donor–acceptor spatial separation R, and will have a series of discrete values depending on the lattice position of the centres. The energy of the transitions is given by

$$\Delta E = E_G - (E_D + E_A) + e^2/\varepsilon R \qquad (2.5)$$

where E_D and E_A are, respectively, the donor and acceptor binding energies and ε is the relative permittivity. This gives rise to a line spectrum with the wavelengths of the lines separated by a few angströms. Although it may not be possible to resolve this spectrum, a donor–acceptor pair recombination may also be identified by observing the change in the spectrum as the emission decays after the excitation is removed. In this situation the recombination at close centres has the shorter time constant, and the spectrum thus shifts to lower energies or longer wavelengths as the decay proceeds. In Figure 2.2, the diagonal line representing the recombination implies that the centres involved have a spatial separation.

[iv] *Recombination via deep levels.* In this case, the photon energy is considerably smaller than the band-gap energy, and the emission wavelength may be so long as to lie outside the region of the spectrum which is of interest. Obviously, the wider the band gap, the greater the likelihood that this type of transition will be of practical interest. With gallium phosphide, a transition of this type gives rise to the red electroluminescent radiation that is commonly observed in this material.

[v] *Exciton transitions.* It has so far been assumed that the free electron or hole taking part in a recombination process occupies one of the continuum of states in the conduction or valence band of the crystal and is not localized in space. In some circumstances however, an electron and a hole can interact to form a state whose energy is less than that of the free electron and hole. This system is called an exciton, and is analogous to an excited electronic state of an atom. In this situation one must consider that the electron and the hole have a spatial association. Furthermore, if one of these charges is localized at a centre, the exciton will be bound to this centre.

For example, a free electron can become associated with a positively charged centre to form a bound exciton. This exciton can then decay with the emission of radiation. A significant feature of this type of transition is that it takes place at a localized centre which can play a part in conserving momentum in the transition. It will be seen later that this can be important in an indirect band-gap material such as gallium phosphide.

2.2.2 *Non-radiative recombination processes*

Recombination processes in which the energy released does not appear as a photon are also possible. These non-radiative transitions are of considerable importance, as they will compete with the desired radiative transitions. Unfortunately, our understanding of many of these types of transition is severely limited, but it will be sufficient for our present purposes to illustrate two possible non-radiative transitions (Figure 2.3).

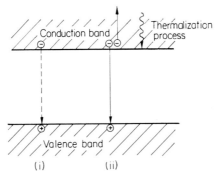

FIGURE 2.3. Non-radiative transitions:
(i) multiphonon and (ii) Auger

[i] *Multiphonon transitions.* An electron may lose energy by the production of a number of quanta of lattice vibrations (phonons). The energy released will thus raise the temperature of the lattice.

[ii] *Auger recombination.* It is possible for an electron to recombine with a hole and give up the excess energy to a second conduction-band electron. The second electron will thus be raised to a higher energy state in the conduction band and will subsequently lose energy by a thermalization process. Since the second electron stays within the continuum of states in the conduction band, no defined energy transitions are involved in the thermalization process and Auger recombination is thus a non-radiative process. Since it involves the interaction of two electrons and a hole, it will become of increasing importance when the concentration of excess electrons is high.

An Auger process can also involve two holes and an electron, and this will be important when the concentration of holes is high.

2.2.3 *Direct and indirect band-gap semiconductors*

From the discussion of the previous section it can be seen that, in the recombination of an electron with a hole, radiative and non-radiative processes are in competition. It is thus necessary to consider the conditions that favour the radiative processes, and at this point it is appropriate to give somewhat closer attention to the band structure of semiconductors. This will give guidance as to which semiconductors might be expected to exhibit a high probability of a radiative transition occurring and a low probability of a non-radiative transition occurring.

On the basis of their band structure, semiconductors can be divided into two classes which are known, respectively, as direct and indirect band-gap materials. The difference between these two classes is illustrated in Figure 2.4. This figure shows the energy E of an electron or hole as a function of momentum $\mathbf{k}$. Since momentum is a vector quantity, the exact details of this plot will depend on the direction of $\mathbf{k}$, but this detail need not be considered here.

The important feature to note in the band structure of the direct band-gap material [Figure 2.4(a)] is that the minimum in the conduction band and the maximum in the valence band occur at the same value of $\mathbf{k}$. This means that, when an electron in the conduction band recombines with a hole in the valence band and a photon is emitted, the total momentum of the system will be conserved. Neglecting the very small amount of momentum carried by the photon, such a transition can be represented by a vertical line on an E–$\mathbf{k}$ plot. A transition of this sort, involving only two particles, will have a high probability.

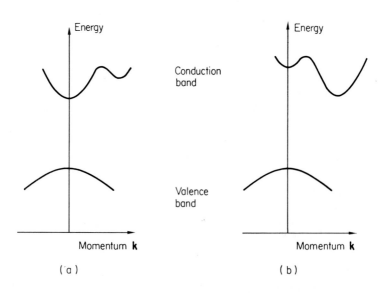

FIGURE 2.4. Energy against momentum for (a) a direct band-gap semiconductor and (b) an indirect band-gap semiconductor

In an indirect band-gap material [Figure 2.4(b)], the top of the valence band and the minimum of the conduction band do not occur at the same value of **k**. In this case the recombination of a hole with an electron with the emission of a photon would not conserve momentum. Since, in any system, momentum must be conserved, the difference in momentum must be accommodated by the crystal lattice, where vibrational states are quantized as phonons. It is thus possible to regard the recombination of an electron with a hole in an indirect band-gap material as a transition involving an additional particle—the phonon—and the transition has a higher order than the equivalent direct process. Qualitatively it can thus be seen that the electron–hole recombination process in an indirect band-gap material will have a lower probability than the recombination occurring in a direct band-gap material. Arguments such as these can be put on a quantitative basis, and it has been estimated[3] that the transition probabilities in the two cases can differ by a factor of the order of 10^7.

It is also of interest to consider the possibility that, in an indirect band-gap material, recombination might occur from a higher-energy direct conduction-band minimum. If it is assumed that the injected or excited electrons are in equilibrium within the conduction band, even if the conduction and valence bands are not in equilibrium, one can calculate the population of the higher conduction-band minima. Since the thermalization time (the time for electrons to reach equilibrium within a band) is very short compared with the recombination times, this is a reasonable assumption. On this basis it can be shown that, for all cases of practical interest, where the upper minima have an energy that is appreciably greater than that at the bottom of the conduction band, the increased probability of a direct recombination is more than countered by the small number of electrons available for such a transition. Hence the prospect of obtaining radiation from a recombination process involving the direct point in the conduction band of an indirect material is slight, and the only band-to-band processes which need to be considered are those involving the lowest levels of the conduction band.

It has been shown that, in an electroluminescent device, radiative and non-radiative processes will be in competition with each other. Such competition will be more significant when the radiative process has a low probability, as in an indirect semiconductor, and the conclusion thus reached is that a direct band-gap material is more likely to give rise to efficient electroluminescent devices. In principle, it might be possible to alleviate this restriction on indirect band-gap material by purifying the material so that non-radiative recombination processes become insignificant. Alternatively, it might be possible to control the properties of the material by the introduction of appropriate impurities so that the basic momentum-conservation rules are, effectively, relaxed. In this way the radiative processes could have an increased probability and dominate the recombination process.

Although such an approach may not at first sight seem encouraging, it will be seen, particularly in connection with gallium phosphide devices, that it is in fact a viable approach, and that useful devices can be made in this way.

2.3   Excitation mechanisms

There are several ways in which a semiconductor can be excited so that the number of electrons in the conduction band is in excess of that required for thermal equilibrium. For example, if a photon of an appropriate energy is absorbed in p-type material, an electron will be raised to the conduction band, leaving a hole in the valence band. Similarly, the system can be excited by a beam of high-energy electrons.

Another potentially useful mechanism makes use of an avalanche process, in which a carrier in the semiconductor material is accelerated by a high electric field. The high-energy or 'hot' carrier which is produced then loses its energy by exciting electrons from the valence band to the conduction band, forming electron–hole pairs. This process will not be very efficient; by considering the conservation of energy and momentum, it can be shown that the hot electron will, on average, have to possess an energy of about $3E$ for the production of an electron–hole pair with a potential energy E.

The luminescence produced as a result of photon excitation (photo-luminescence) and that produced by electron excitation (cathodoluminescence) are useful tools in the investigation of semiconductor materials and can give information about the recombination transitions which occur in a semiconductor. However, for practical devices, some form of diode, operating at a voltage approximately equal to the band-gap potential, is highly desirable. In this context two systems merit further consideration. The first of these is the p–n junction, which at the present time has found almost universal application. Secondly, it is necessary to consider in outline the possibility of utilizing a metal–semiconductor junction. Although this latter approach has not yet proved to be efficient, it is an approach which, if successful, might offer several advantages over the p–n junction.

2.3.1   The p–n junction

The band structure of a p–n junction under zero and forward bias is shown in Figure 2.5. Under zero bias the system is in equilibrium, and the Fermi level, which describes the population of the energy bands, has a constant value. There will thus exist, across the junction, a potential barrier which will be approximately equal to the band gap of the material. When the junction is forward biased, the potential barrier is lowered and a current will flow through the junction. This current flow can be due to a number of processes, but usually only one gives rise to an excitation which results in a radiative recombination process. These processes are illustrated in Figure 2.5.

[i] *Diffusion current.* When the junction barrier is lowered, electrons from the *n*-type side of the material will diffuse across the junction to the *p* side and recombine with the majority carriers (holes). On average, the minority

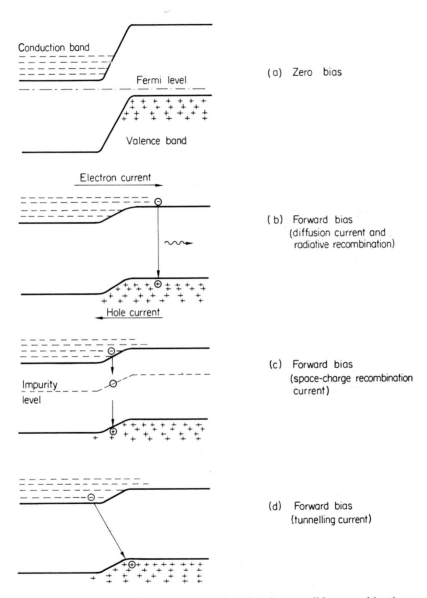

FIGURE 2.5. Energy levels of a *p–n* junction showing possible recombination processes and current flow (for clarity, donor and acceptor states are not shown)

electrons will have a lifetime τ_e and will travel the diffusion length L_e before recombining. The diffusion length is given by

$$L_e = \sqrt{(D_e \tau_e)} \qquad (2.6)$$

where $$D_e = \mu_e \frac{kT}{e} \quad \text{(Einstein's relationship)} \qquad (2.7)$$

D_e is the electron diffusion coefficient and μ_e is the electron mobility.

In the same way, holes will diffuse across the junction into the n-type material, and both the diffusion currents will contribute additively to the total junction current. The standard theory of the p–n junction shows that the current–voltage characteristic of the p–n junction is

$$J = e \left(\frac{D_e n_p}{L_e} + \frac{D_h p_n}{L_h} \right) \left(\exp \frac{eV}{kT} - 1 \right) \qquad (2.8)$$

where J is the current density at an applied voltage V, n_p is the equilibrium concentration of minority electrons in the p-type material, p_n is the equilibrium concentration of minority holes in the n-type material and the subscripts e and h specify the diffusion coefficient D and minority-carrier diffusion length L of electrons and holes.

In equation (2.8), the first term refers to the electron contribution to the total current and the second term to the hole current. Often only one of these components (usually the electron current) gives rise to radiative recombination, and the junction must be designed to make this component dominant.

The ratio of electron to hole currents is known as the injection ratio of the junction and is given by

$$\frac{J_e}{J_h} = \frac{D_e n_p}{L_e} \times \frac{L_h}{D_n p_n} = \frac{L_h \sigma_n}{L_e \sigma_p} \qquad (2.9)$$

where σ_n and σ_p are the conductivities of the n and p regions.

As a rough approximation, this ratio is given by the ratio of the conductivities of the two materials. The electron current will thus be favoured if the n region has a high carrier concentration and the electrons have a high mobility

In all practical cases, electroluminescent diodes are operated at a forward bias such that $eV \gg kT$ and the diode characteristic can be represented by

$$J = A \exp \frac{eV}{kT}. \qquad (2.10)$$

[ii] *Space-charge recombination current.* The electric field that exists in a p–n junction is associated with a net charge distribution, and the region over which this exists is known as the space-charge region of the junction. The simple diffusion-current analysis of the junction assumes that no

minority-carrier recombination occurs in this region, although this approximation is often not valid. If a number of deep energy levels, near the centre of the band gap, exist in this region, recombination of minority carriers will occur via these levels, giving rise to a current component known as the space-charge recombination current. The density of this current is given by the expression

$$J = B \exp\frac{eV}{nkT} \qquad (2.11)$$

where B is a constant and n lies between 1 and 2.

In the simplest cases, the coefficient n is approximately equal to 2. This component of current is in parallel with the diffusion current, but does not give rise to the required radiative process. Hence, in practical devices, attempts are made to make this current component as small as possible by minimizing the number of unwanted deep energy levels in the semiconductor material. This implies that great care must be taken to exclude unwanted impurities from the system.

[iii] *Tunnel current.* If the junction is narrow and the material on either side highly doped, a component of current due to tunneling can be observed. In this mechanism, electrons from the n side of the junction recombine directly with holes on the p side of the junction by tunneling through the forbidden band gap. The tunnel current is given by

$$J = C \exp\alpha V \qquad (2.12)$$

where C and α are constants. It should be noted that this expression does not involve a function of eV/kT, and this serves to differentiate experimentally a tunnel current from other mechanisms. Since this tunnel current does not inject minority carriers, it does not contribute to a radiative recombination process.

In most practical diodes the tunnel current is small, and it is sufficient to consider only the diffusion and space-charge recombination currents. Thus the total diode current I is given by

$$I = A'\exp\frac{eV}{kT} + B'\exp\frac{eV}{2kT}. \qquad (2.13)$$

The space-charge recombination current may dominate at low values of bias, but the diffusion current will always dominate at sufficiently high bias.

The bias which occurs in equation (2.13) is the voltage which appears across the junction. If the diode has an appreciable resistance R, a voltage IR will be dropped across this resistance, and equation (2.13) must be modified to

$$I = A'\exp\frac{e(V-IR)}{kT} + B'\exp\frac{e(V-IR)}{2kT} \qquad (2.14)$$

where V is the voltage applied to the device.

If the current–voltage characteristic of a diode is examined, it should be possible to distinguish three regions (Figure 2.6):

(a) Low current:

$$I \sim \exp\frac{eV}{2kT}.$$
(2.15)

(b) Intermediate current:

$$I \sim \exp\frac{eV}{kT}.$$
(2.16)

(c) High current:

$$I \sim V/R.$$
(2.17)

In practice, it may be difficult to identify these three regions, as any of them may be clearly defined over only a limited region of the characteristic.

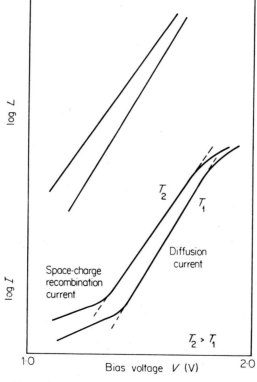

FIGURE 2.6. Current–voltage and output–voltage
characteristics of a *p–n* junction

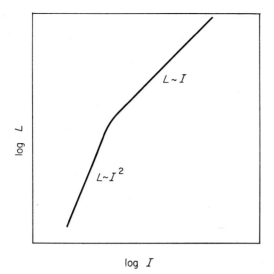

FIGURE 2.7. Output against current for an electroluminescent diode

The radiative output from the device is due to the diffusion current and is thus given by an expression of the form

$$L = C \, \exp\frac{eV}{kT} \qquad (2.18)$$

where C is a constant whose value depends on the efficiency of radiative recombination and may vary with temperature.

Often the device characteristic which is more important is that relating output to current and this will, in general, show two regions, depending on the dominant current mechanism. For the low-current region, in which space-charge recombination dominates the current flow, equations (2.15) and (2.18) give

$$L \sim I^2 \qquad (2.19)$$

and, for the intermediate- and high-current regions

$$L \sim I. \qquad (2.20)$$

These characteristics are illustrated in Figure 2.7.

The curves shown in Figures 2.6 and 2.7 represent the idealized situation corresponding to equation (2.13). In practice, experimental curves are more accurately represented by an equation of the form

$$I = A'' \exp\frac{eV}{akt} + B'' \exp\frac{eV}{bkT} \qquad (2.21)$$

where the values of *a* and *b* may depart appreciably from the integer values $a = 1$ and $b = 2$, representing diffusion and space-charge recombination currents, respectively. It may then be difficult to identify the various regions of the diode characteristics.

2.3.2 Metal–semiconductor junctions

Metal–semiconductor junctions or Schottky barriers have been studied for many years, but it is only recently, with the aid of modern technology, that reproducible and near-ideal structures have been produced. Such structures might give a means of injecting minority carriers into a semiconductor, and thus give rise to radiative recombination. At present, considerable effort is being devoted to this problem, although, as yet, no practical devices have been developed. However, the advantages of this approach are such that it should not be neglected. Figure 2.8 shows the band structure of an idealized situation in which a metal is brought into contact with a semiconductor. In this idealized situation, surface states on the semiconductor have been neglected, and there is no oxide film between the metal and the semiconductor. In this figure ϕ_m and ϕ_s are the work functions

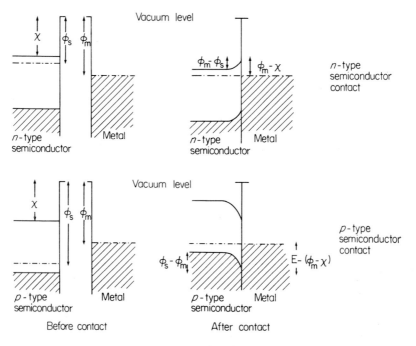

FIGURE 2.8. Schematic diagram of the band structure of a semiconductor–metal contact (neglecting surface states)

of the metal and semiconductor, respectively, and χ is the electron affinity of the semiconductor.

When contact is made between the metal and the semiconductor, the Fermi level in the system must be continuous, and the semiconductor bands are 'bent' as shown, giving rise to a potential barrier across the junction with lines of force terminating on charges in the space-charge region of the semiconductor. The barrier height seen by an electron in the n-type semiconductor is given by

$$V_s = \phi_m - \phi_s. \tag{2.22}$$

The barrier height seen by an electron in the metal is given by

$$V_m = \phi_m - \psi. \tag{2.23}$$

Both of these barrier heights depend directly on the work function of the metal.

When the junction is forward biased, a current flow due to several components will be established. For the present discussion, taking n-type material as an example, it is only necessary to consider the current of electrons (majority carriers) into the metal and the injection of holes (minority carriers) into the semiconductor. The standard treatment of the Schottky barrier predicts that the electron current density J_e produced by a bias potential V will be given by

$$J_e = A \left(\exp\frac{eV}{kT} - 1 \right) \tag{2.24}$$

giving rise to a diode characteristic. The ratio of the electron current to the hole injection current can also be predicted. This ratio, which includes an exponential function of the metal work function, the semiconductor band gap and the semiconductor electron affinity, is given by an expression of the form

$$\frac{J_e}{J_h} \sim \exp \left(\frac{E_G - \phi_m + \psi}{kT} \right). \tag{2.25}$$

Hole injection will thus be favoured in a system involving a low-energy-gap semiconductor and a high-work-function metal. In a similar way, electron injection into a p-type semiconductor will be favoured by a metal with a low work function.

This simple model thus suggests that the required minority carrier injection might be obtained by a suitable choice of metallic contact and work function. In practice, the situation is not as simple as the idealized case considered here might suggest. This is due to the existence of surface states on the semiconductor and interfacial layers between the metal and the semiconductor.

The effect of these surface states is to bend the semiconductor bands

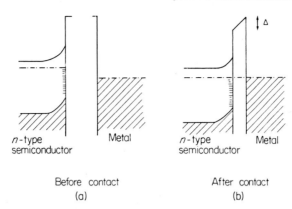

| | Before contact | After contact |
| | (a) | (b) |

Before contact   After contact
(a)   (b)

FIGURE 2.9. Energy-band diagram of a semiconductor–
metal contact, showing the effect of surface states

near the surface of the material as shown in Figure 2.9(a). When the semiconductor is brought into contact with a metal, the situation is as shown
in Figure 2.9(b). Lines of force can terminate on charges in the surface
states, and no further band bending occurs. A finite voltage is dropped
across the interface, and the potential-barrier expressions given previously
do not hold. The dependence of the barrier height on the metal work function
is reduced and it can be shown that the greater the density of surface states,
the less dependent is the barrier height on the work function. As an illus-
tration of this, some typical values of barrier height for a range of metals
on gallium arsenide are given in Table 2.1. In this example, the metal work
function varies by 1·3 V, while the barrier height varies by less than 0·2 V.

Hence it can be seen that the effect of surface states is that one of the
degrees of freedom in designing a metal–semiconductor barrier is curtailed
and, in practice, it becomes extremely difficult to obtain useful minority-
carrier injection currents in such a structure.

TABLE 2.1. Barrier heights on gallium arsenide.

Metal	Work function (V)	Barrier height (V)	
		n type	p type
Au	4·8	0·95	0·48
Pt	5·2	0·94	0·48
Ag	4·5	0·93	0·44
Al	4·2	0·80	0·63
Be	3·9	0·82	

2.4   Electroluminescent materials

It has been shown in the previous sections that there are two basic properties required of a semiconductor material which is to be used as the basis of an electroluminescent diode. The first of these requirements is that the material should exhibit suitable luminescence properties; i.e. that it should show a high probability of a radiative transition occurring in the required region of the spectrum. It has also been shown that it is to be expected that a direct band-gap material would exhibit favourable recombination properties, although it may also be possible to obtain high recombination probabilities in an indirect material under certain circumstances. The second requirement is that it should be possible to construct a *p–n* junction in the material. The semiconductor materials which meet these requirements will now be considered in general terms, while a detailed discussion of experimental results is reserved for Chapter 4.

2.4.1   *The group IV semiconductors*

The group IV elemental semiconductors germanium and silicon have both been extensively developed for use in transistors and other devices. Unfortunately both these materials have indirect band gaps (germanium 0·7 eV, silicon 1·1 eV). Hence, although it has been possible to observe radiative recombination from both of these materials, this is not an efficient process, and the radiation is in the near-infrared region of the spectrum. Thus no practical electroluminescent diodes based on germanium or silicon have been developed.

Silicon carbide (SiC) is a semiconductor material that can exist in a number of forms or polytypes of different crystal structure. It is an indirect gap material, like germanium and silicon, and has a band gap of 2–3 eV, depending on the polytype. The polytype which is known as 6H has a band gap of 3 eV and radiation with a corresponding photon energy has a wavelength of approximately 4000 Å, which is in the blue region of the spectrum. As is to be expected, diodes at this wavelength have a low efficiency. Impurities can be introduced into silicon carbide and, with a suitable choice of impurities, a range of radiative transitions can be observed, giving diodes emitting in the green, yellow or red regions of the spectrum. Again, the efficiency of devices based on these transitions is low.

2.4.2   *The III–V compound semiconductors*

A section of the periodic table, showing some of the elements in groups II–VI, is shown in Table 2.2. The elemental semiconductors silicon and germanium are found in group IV. Besides these and other elements, many chemical compounds are semiconductors. The most widely studied are the III–V compound semiconductors,[4] which are formed from equal atomic

B

TABLE 2.2. Elements in a section of the periodic table.

II	III	Group IV	V	VI
Beryllium Be	Boron B	Carbon C	Nitrogen N	Oxygen O
Magnesium Mg	Aluminium Al	Silicon Si	Phosphorus P	Sulphur S
Zinc Zn	Gallium Ga	Germanium Ge	Arsenic As	Selenium Se
Cadmium Cd	Indium In	Tin Sn	Antimony Sb	Tellurium Te

proportions of one of the group III elements aluminium, gallium or indium and one of the group V elements phosphorus, arsenic or antimony. These nine compounds all have semiconducting properties, and some of the important features of the band structure of these materials are given in Table 2.3. It can be seen that the band gap increases with decreasing atomic weight of both the group III and the group V elements. The materials in

TABLE 2.3. Band structure of III–V compounds. This table gives the position of the [000] and [100] conduction-band minima with respect to the valence band, and indicates the direct or indirect nature of the band gap. Parameters are room-temperature values in electronvolts.

Group III element	Group V element		
	Phosphorus	Arsenic	Antimony
Aluminium	AlP ? : 2·4 Indirect	AlAs 2·83 : 2·13 Indirect	AlSb 1·62 : 1·53 Indirect
Gallium	GaP 2·65 : 2·25 Indirect	GaAs 1·44 : 1·86 Direct	GaSb 0·68 : 2·10 Direct
Indium	InP 1·3 : 2·1 Direct	InAs 0·35 : 2·0 Direct	InSb 0·17 : 2·0 Direct

this table can be divided into direct and indirect band-gap materials with the higher band-gap materials having an indirect band gap. Table 2.3 also shows the height of the [000] and [100] conduction-band minima above the top of the valence band. For a direct-gap material, the [000] minimum is lower than the [100] minimum.

From Table 2.3, it can be seen that gallium arsenide is the III–V compound which has the largest direct band gap. Consequently gallium arsenide is the basis of a number of efficient electroluminescent devices. These emit with a photon energy of 1·4 eV, corresponding to a wavelength of 0·9 μm, in the near-infrared region of the spectrum. The other direct band-gap materials in Table 2.3 can also give rise to efficient electroluminescent diodes, but, since the wavelength of their emission is in a region of the spectrum which is of less practical importance, less attention has been paid to them.

Considering the indirect-gap materials in Table 2.3, it might, for example, be assumed that gallium phosphide would not be a worthwhile material for electroluminescent diodes. However, as will be seen in a later chapter, early experimental work showed that gallium phosphide could exhibit high electroluminescent efficiencies if doped in suitable ways,[5] and extensive work on gallium phosphide has given rise to a range of important devices.

Although the term III–V compounds is often restricted to the nine compounds referred to, it should, of course, include the compounds containing boron or nitrogen. Although these materials present extreme technological difficulties, gallium nitride has been shown to have a band gap of 3·4 eV, which would correspond to a photon in the ultraviolet region of the spectrum. However, by searching for suitable impurity transitions, it may be anticipated that a range of photon energies throughout the visible spectrum might be obtained. This approach is, at present, unproven but has aroused considerable interest.

2.4.3 The $III^A_x III^B_{1-x} V$ and $III V^A_x V^B_{1-x}$ ternary materials

The search for direct band-gap semiconductors with a band gap greater than that of gallium arsenide leads to a consideration of materials of the type $III^A_x III^B_{1-x} V$ and $III V^A_x V^B_{1-x}$ which are formed from two of the compounds in one row or column in Table 2.3. The III–V semiconductor compounds are miscible in all proportions, and many of the properties of these materials can be deduced by interpolating between the properties of the constituent binary compounds. In particular, the essential features of the band structure of the semiconductors which are of interest as electroluminescent materials can be obtained by interpolating the data given in Table 2.3.

The band structures of a series of compounds formed from different proportions of one direct and one indirect band-gap material shows a transition between the direct and indirect structures. Thus, by alloying a direct-gap semiconductor with an indirect-gap semiconductor (with a wider band gap), it will be possible to obtain a material with a direct band gap that is wider than that of the first material.

To obtain transitions giving rise to visible radiation, the energy gap of the semiconductor must be 2 eV or more. Thus the III–V ternary materials which must be considered are InP–GaP, GaAs–GaP, GaAs–AlAs, InP–AlP, GaSb–GaP and InAs–AlAs. The band structure of these materials is shown

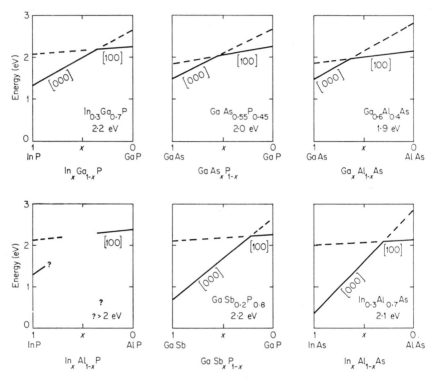

FIGURE 2.10. Band-structure parameters of III–V ternary materials, giving composition and band gap of direct–indirect transition materials at 300 K

in Figure 2.10, and it can be seen that all of them show a transition between the direct and indirect structures at a band gap of about 2 eV.

Consider, for example, the gallium phosphide–gallium arsenide system. Gallium arsenide has a direct band gap of 1·44 eV, with a subsidiary minimum 0·42 eV higher, at 1·86 eV. Gallium phosphide has an indirect band gap of 2·25 eV, with a second minimum 0·4 eV higher, at 2·65 eV. A range of alloys can be made from these two components and the alloy material changes from a direct to an indirect band structure at a composition of $GaAs_{0.55}P_{0.45}$. At this composition the band gap is 2·0 eV, corresponding to a photon wavelength of 6300 Å.

Although the band-gap data for other materials is not so accurately known as for the GaAs–GaP system, similar estimates can be made for other materials of interest. These calculations suggest that the InP–GaP system will have a maximum direct band gap at a composition $In_{0.3}Ga_{0.7}P$. At this composition, the band gap (2·2 eV) corresponds to a photon wavelength of 5700 Å. Similarly, the GaAs–AlAs system will have a direct band gap of about 2 eV at the composition $Ga_{0.6}Al_{0.4}As$, and, although the band gap is no larger than that found in the GaAs–GaP system, gallium–aluminium arsenide alloys have some technological advantages to commend them.

All the III–V binary compounds and ternary alloys which are of interest can be made both p type and n type, and thus give a range of materials which are of interest in electroluminescent devices.

2.4.4 The II–VI compounds

The compounds formed from equal atomic proportions of one of the group II elements zinc, cadmium or mercury and one of the group VI elements oxygen, sulphur, selenium or tellurium form a series of semiconducting materials. The known band-gap energies of these materials are shown in Table 2.4. These band-gap energies increase with the decreasing atomic weight of the elements, from HgSe and HgTe, which are semimetals, to ZnS, with a band-gap energy greater than 3 eV. All these semiconducting compounds have a direct band gap, and can thus be expected to have

TABLE 2.4. Band gap and conductivity type of the II–VI compounds. Band-gap energies are room-temperature values in electronvolts.

Group II element	Group VI element		
	Sulphur	Selenium	Tellurium
Zinc	ZnS 3·6	ZnSe 2·6 n type	ZnTe 2·2 p type
Cadmium	CdS 2·6 n type	CdSe 1·7	CdTe 1·6 Amphoteric
Mercury	HgS 2·1	HgSe Semimetal	HgTe Semimetal

efficient luminescent properties. This is in fact the case and the II–VI compounds, in particular ZnS, give rise to very efficient phosphors. Unfortunately these compounds have limited conducting properties. The ionic component of the crystal binding increases with decreasing atomic weight and, associated with this, it becomes increasingly difficult to produce both p- and n-type conductivity. This is owing to a self-compensating mechanism by which any donor or acceptor elements that are added to the material are automatically compensated by electrically active lattice defects. In fact, of the compounds shown in Table 2.4, only CdTe is amphoteric (i.e. can be doped to be both p and n type). ZnSe and ZnTe can only be doped to low-resistivity n and p type, respectively.

An obvious approach to this problem is to study the properties of suitable alloys of the II–VI compounds. For example, it has been shown that the alloy $ZnSe_xTe_{1-x}$ is amphoteric over the range given by $0{\cdot}4 < x < 0{\cdot}5$, and p–n junctions can be made in this material.

Alternatively, it is possible to make heterojunctions in which the p and n sides of the junction are of different materials. Here it may be difficult to decide whether a true heterojunction has been made, or whether an amphoteric alloy has been formed between the materials on either side of the junction.

The other approach, which shows some promise with the II–VI compounds, lies in the use of metal–semiconductor or metal–insulator–semiconductor contacts. Although this approach has not, as yet, yielded any practical devices, it will be considered in more detail in the section devoted to the II–VI compounds in Chapter 4, as it may well be that future work in this field will be rewarding.

2.5 p–n junction lasers

Any laser has several basic requirements and in this section it will be shown how these can be met in a semiconductor.

The first requirement is for a medium which can be excited so that it will amplify, by stimulated emission, photons of the appropriate wavelength. A semiconductor can provide a suitable medium when a high density of excess minority carriers is injected into it.

The second requirement is for a means of excitation whereby the population of energy levels in the medium is perturbed from its thermal-equilibrium distribution. In a semiconductor this is conveniently done by using a p–n junction to inject excess electrons into p-type material.

The third requirement is for a resonant cavity so that the amplification will produce a state of oscillation. The resonant cavity will endow the laser with characteristic spectral properties and give the beam of radiation which emerges the high directionality which is an important feature of all lasers.

2.5.1 Stimulated emission in semiconductors

Although there is no fundamental reason why laser action should not be observed in the indirect band-gap semiconductors, the radiative transition probabilities in these materials are low, and consequently laser action has only been observed in direct band-gap materials. In discussing the theory of laser action in semiconductors, it will thus be assumed that only direct band-gap semiconductors need to be considered and, furthermore, that the only transitions that are relevant are radiative transitions between the valence and conduction bands or states close to these bands. It will also be convenient to consider the processes which occur in p-type materials, although an analogous consideration will apply to n-type material.

The possibility of obtaining stimulated emission in a semiconductor was first considered by Basov, Krokhin and Popov,[6] and by Bernard and Durraforg.[7] These authors derived, using the concept of a quasi-Fermi level, the condition under which a photon flux in a semiconducting medium would be amplified rather than attenuated, as would occur if the medium were in thermal equilibrium.

When a semiconductor is in thermal equilibrium, the probability of occupation of any state of energy E is given by the Fermi–Dirac function

$$f = \frac{1}{1 + \exp\dfrac{E - F_0}{kT}} \qquad (2.26)$$

where F_0 is the Fermi level. If the system is not in thermal equilibrium, it can be shown that the population of either of the bands can be described by the same function, provided that two quasi-Fermi levels are defined: F_c relating to the conduction band and F_v to the valence band (Figure 2.11).

When a photon flux is incident on a semiconductor, the rates of stimulated emission $R_{e,st}$ and stimulated absorption $R_{a,st}$ are given by

$$R_{e,st} = A f_e (1 - f_v) \qquad (2.27)$$

$$R_{a,st} = A f_v (1 - f_e) \qquad (2.28)$$

where the constant A involves the density of states in the valence and conduction bands, the transition probability and the photon flux. f_e and f_v are the occupation probabilities of states of energy E_c and E_v in the conduction and valence bands such that $E_c - E_v = h\nu$, the photon energy of radiation.

For the photon flux to be amplified, the rate of stimulated emission must exceed the rate of absorption.

Thus
$$f_e (1 - f_v) > f_v (1 - f_e)$$

or
$$f_e > f_v. \qquad (2.29)$$

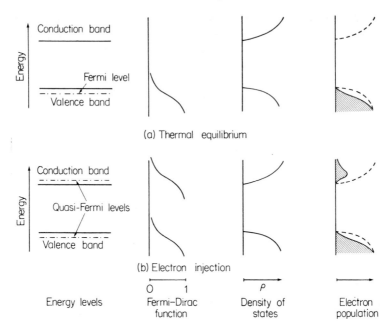

FIGURE 2.11. The Fermi–Dirac function, density of states and electron population of a *p*-type semiconductor (a) in thermal equilibrium and (b) with electrons injected into the material

When this condition is fulfilled, the system has an inverted population in which the probability of a higher-energy state being occupied is greater than that of the lower-energy state being occupied. By substituting equation (2.26) in equation (2.29), one obtains

$$F_c - F_0 > h\nu. \tag{2.30}$$

Although this condition does not enable one to calculate the magnitude of the gain in a system, it provides a convenient way of discussing the possibility of laser action, particularly in *p–n* junction structures.

Much of the understanding of laser action in semiconductors is due to the work of Lasher and Stern,[8] and a brief outline of their calculation of the gain due to stimulated emission in a semiconductor is given here.

It can be shown that the rates of spontaneous and stimulated emission per unit volume for transitions between two bands are given by

$$R_{sp} = \frac{4\mu e^2 h\nu}{m^2 \hbar^2 c^3} \sum |M|^2 [f_c(1 - f_v)] \tag{2.31}$$

$$NR_{st} = \frac{4\mu e^2 h\nu N}{m^2 \hbar^2 c^3} \sum |M|^2 [f_c(1 - f_v) - f_v(1 - f_c)] \tag{2.32}$$

where M is the matrix element for the transition, N is the photon density, μ is the refractive index of the medium and $h\nu$ is the photon energy. The summations are taken over all pairs of states with an energy difference $h\nu$.

The stimulated-emission term is the net rate of emission and absorption and is given by

$$R_{st} = R_{e,st} - R_{a,st}.$$

The total rate at which photons of energy E are emitted into the system is

$$R_e = R_{sp} + NR_{st}. \tag{2.33}$$

Putting $F_c - F_v = \Delta F$, and using the usual expressions for the Fermi–Dirac functions f_c and f_v:

$$R_{st} = R_{sp}\left[1 - \exp\frac{E - \Delta F}{kT}\right]. \tag{2.34}$$

For emission to exceed absorption, R_{st} is positive and $\Delta F > E$, as in equation (2.30).

It is sometimes useful to be able to relate the emission functions defined in equations (2.31) and (2.32) to a property of the semiconductor, such as its absorption coefficient α, that is amenable to experimental investigation. It can be shown that these are related by the expression[9]

$$\alpha(h\nu) = \frac{\pi^2 c^2 \hbar^2}{\mu^2 (h\nu)^2} R_{st}(h\nu). \tag{2.35}$$

Here again, the value of α depends on the quasi-Fermi levels, and the value obtained for a material in thermal equilibrium is given by putting $F_c = F_v = F_0$.

To make further use of equations (2.31) and (2.32), some assumptions must be made about the matrix elements M. In relatively pure material, transitions would be allowed only between states of the same wave vector; but this will not necessarily hold in heavily doped material, such as is used in lasers. Going to the opposite extreme, Lasher and Stern assumed that the matrix element would be the same for all transitions. Hence equations (2.31) and (2.32) become

$$R_{sp}(h\nu) = C\int \rho_c(E) \cdot \rho_v(E - h\nu) \cdot f_c(E)[1 - f_v(E - h\nu)]dE \tag{2.36}$$

$$R_{st}(h\nu) = C\int \rho_c(E) \cdot \rho_v(E - h\nu)[f_c(E) - f_v(E - h\nu)]dE \tag{2.37}$$

where C is a constant and ρ_v and ρ_c are the densities of states in the valence and conduction bands. The integration extends from $E = 0$ to $E = h\nu - E_G$ to include all pairs of states in which the energy separation is $h\nu$, as shown in Figure 2.12.

Further assumptions are now needed regarding the density-of-states

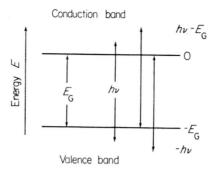

FIGURE 2.12. Integration limits required to evaluate spontaneous- and stimulated-emission functions

function. The simplest assumption is that the valence and conduction bands are parabolic with

$$\rho_c \sim E^{\frac{1}{2}} \tag{2.38}$$

$$\rho_v \sim (h\nu - E - E_G)^{\frac{1}{2}}. \tag{2.39}$$

The spontaneous- and stimulated-emission functions calculated on this basis are shown in Figure 2.13. It should be remembered that these functions have, as a parameter, the positions of the quasi-Fermi levels which, in turn,

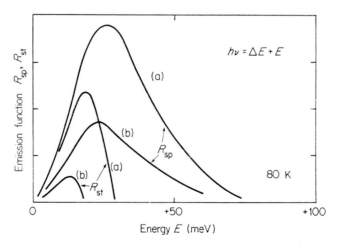

FIGURE 2.13. Spontaneous- and stimulated-emission functions (a) injection level = 2×10^{17} cm^{-3}, quasi-Fermi level = + 15 meV and (b) injection level = 10^{17} cm^{-3}, quasi-Fermi level = +5 meV. [After G. J. Lasher and F. Stern, *Phys. Rev.*, **133**, A553 (1964) and reproduced by permission.]

are determined by the concentration of electrons and holes in the system. The curves shown in Figure 2.13 are calculated for *p*-type gallium arsenide at 80 K with a hole concentration of $3 \times 10^{18} \text{cm}^{-3}$. The injected-electron concentrations for these two curves are 10^{17} and $2 \times 10^{17} \text{cm}^{-3}$, corresponding, respectively, to quasi-Fermi levels 5 MeV and 15 meV above the bottom of the conduction band.

Although the exact shape of the curves shown in Figure 2.13 depends on the details of the model used, the general features remain valid and are sufficiently important to warrant further comment.

The spontaneous-emission curves show the rate of spontaneous emission for photon energies extending upwards from the band-gap energy. These curves thus predict the spectral shape of the spontaneous emission for different injection rates.

The stimulated-emission curves are positive for photon energies over a finite range above the band-gap energy. This corresponds to the stimulated amplification of radiation. At higher photon energies the curves become negative, showing that high-energy radiation will be absorbed. As the injected-electron concentration increases, the peak of the curve moves to higher photon energies and its magnitude increases.

The exact conduction- and valence-band shapes and the momentum rules to be applied to any particular situation have been the subject of much discussion. In particular, the effect of high doping levels has received considerable attention, both theoretically and experimentally. As the doping level is increased, the impurity level will broaden and interact with the associated conduction or valence band until the impurity levels form a tail on the band. These effects will be most pronounced when the impurity ionization energy and the effective mass of a carrier in the band are small.

Various models have been used to describe these effects, and the band tail can be approximated in several ways. For example, the density of states in the band tail as a function of energy E can be represented as an exponential function[10]

$$\rho(E) \sim \exp(-E/E_0) \qquad (2.40)$$

or a Gaussian function[11]

$$\rho(E) \sim \exp[-(E/E_1)^2] \qquad (2.41)$$

where E_0 and E_1 are energy parameters which are characteristic of the band tail. With the aid of these models, it is possible to predict the stimulated-emission function as a function of photon energy, temperature, doping level and minority-carrier concentration.[12,13]

Laser action will occur at the wavelength at which the stimulated emission rate is greatest and will take place when the gain due to stimulated emission is sufficient to overcome the losses in the system. Thus the curve shown in Figure 2.13 can be related to the threshold condition of the laser by plotting the maxima of the stimulated-emission function as a function of the injection level, temperature or material parameters. In this way the effect of these parameters on device performance can be predicted.

The fundamental effect of temperature is to broaden the energy distribution of the injected minority carrier and thus reduce the stimulated emission at any particular wavelength. Thus, as the temperature is increased, the injected minority-carrier density must be increased to achieve laser action. At low temperatures, the gain decreases slowly with temperature and the required injection level is almost constant. At higher temperatures, the gain decreases more rapidly with temperature and, in many cases, the required injection level increases as the cube of temperature.

Studies of the effect of doping level on the stimulated-emission function are hampered by the lack of knowledge of the effect of high doping levels on the system. In general, however, at low temperatures, a high doping level will reduce the optical gain in the system; at higher temperatures the converse is true and a high doping level will reduce the injection level required to achieve laser action. This is in qualitative agreement with the experimental results that will be considered in Chapter 5.

2.5.2   Injection at a p–n junction

Minority-carrier injection and laser action can be achieved by bombarding a semiconductor with a beam of fast electrons. This technique is particularly useful for semiconductors which are not amphoteric, but does not give rise to a practical device. Injection by means of a *p–n* junction is the most useful and viable means of exciting laser action in a semiconductor.

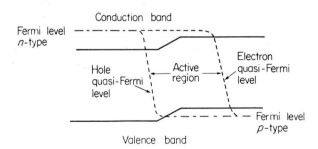

FIGURE 2.14. Energy levels in a highly doped, forward-biased *p–n* junction

The band structure of a *p–n* junction is shown in Figure 2.14. Under forward bias, minority carriers are injected across the junction and the quasi-Fermi levels have the general form shown. Providing that at least one side of the junction is degenerately doped, a high forward bias produces a region in which the quasi-Fermi levels are separated by an energy greater than the band gap of the semiconductor. Thus the condition expressed by equation (2.30) is satisfied and this region has an inverted population which can amplify a photon flux of suitable photon energy.

An exact analysis of the current flow in a laser *p–n* junction is hampered by a lack of detailed knowledge of such properties as minority-carrier lifetime and mobility in the heavily and assymetrically doped structures. The problem can, however, be illustrated by considering a typical gallium arsenide junction in which the *n*-type doping level is $2 \times 10^{18} \text{cm}^{-3}$ and that of the *p* region is $5 \times 10^{18} \text{cm}^{-3}$. The properties of the *n* and *p* regions of such a junction are given in Table 2.5 and have been calculated as discussed in Sections 2.1 and 2.2.

TABLE 2.5. Properties of the *n* and *p* regions of a gallium arsenide laser junction (the values are for a typical GaAs junction at 77 K).

	p type	*n* type
Doping level (cm⁻³)	5×10^{18}	2×10^{18}
Minority carrier	Electron	Hole
Lifetime (s)	10^{-9}	2×10^{-9}
Mobility (cm²/V s)	2000	100
Diffusion coefficient (cm²/s)	50	2
Diffusion length (μm)	1·5	0·7

The ratio of electron to hole currents for this junction is approximately 4 so that 80 per cent. of the current is carried by electrons, and minority-carrier injection extends approximately 1·5 μm into the *p* side of the junction. This is in qualitative agreement with the observed features of gallium arsenide laser junctions in which laser action appears to occur on the *p* side of the junction over a region about 2 μm wide.

2.5.3 *The laser cavity and threshold condition*

It has been shown that minority carriers can be injected across a *p–n* junction with a density that is sufficient to produce an inverted population and thus give rise to stimulated emission. It is now necessary to see how such a junction can be used as a resonant cavity and produce coherent oscillation.

The Fabry–Pérot structure used to produce coherent oscillation is shown in Figure 2.15. The junction region is terminated by polished faces of the semiconductor which are perpendicular to the junction plane and parallel to each other. These faces have a reflectivity R_1 and R_2 and are a distance L apart.

[i] *Threshold condition.* The threshold condition for oscillation of such a cavity is given by considering the net amplification of a photon flux that travels in the junction plane and is reflected at the ends of the cavity. At threshold, the net gain is unity. Thus:

$$R_1 R_2 \exp[g(J) - \alpha] L = 1 \tag{2.42}$$

$$g(J) = \alpha + \frac{1}{L} \ln \frac{1}{\sqrt{(R_1 R_2)}} \tag{2.43}$$

where $g(J)$ is the gain per unit length of the cavity and is a function of the current density J; α represents the cavity losses per unit length and is analogous to an absorption coefficient. This relationship has been investigated experimentally for a wide range of devices. By varying the cavity length or reflectivity, values of the gain $g(J)$ and loss coefficient α can be obtained. The results of these measurements are discussed in more detail in Chapter 5.

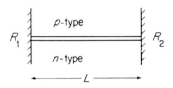

FIGURE 2.15. Section of Fabry–
Pérot laser cavity

[ii] *Gain.* In many situations, it is found that the gain in a laser is a linear function of the current density, so that $g(J)$ can be written as

$$g(J) = \beta J. \tag{2.44}$$

However this relationship is not universally valid and care must be taken not to make this assumption and thus arrive at an erroneous interpretation of experimental results.

[iii] *Cavity losses.* In a semiconductor laser, an electromagnetic wave propagates through a narrow amplifying region which is bounded by absorbing regions. The electromagnetic wave will extend outside the amplifying region and thus be attenuated. Several attempts have been made to analyse this situation. Anderson[14] considered a model which consisted of three layers of different relative permittivity (Figure 2.16). By introducing a complex relative permittivity, this model takes into account variations in both refractive index and absorption in the three regions.

This structure can propogate both transverse electric (TE) and transverse magnetic (TM) modes. It is found that, for the lowest-order modes, in which the electromagnetic field has a maximum at the centre of the active region, the TE and TM modes have a similar attenuation. Thus it is to be expected that the radiation from a laser would have approximately equal probabilities of being polarized either in the plane of the junction or perpendicular to the junction.

If it is assumed that the relative permittivity is constant throughout the structure, the values of attenuation that are predicted are much larger than those that are observed in practical devices. This leads to the conclusion that the wave is confined to some extent to the active regions of the device so that less energy is absorbed in the inactive regions. Such confinement, which is often referred to as 'mode confinement' would be expected if the refractive index of the medium varied in a suitable way across the structure. To account for the observed attenuation, this variation must be about

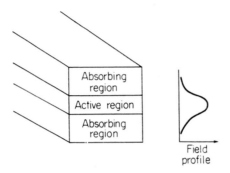

FIGURE 2.16. Dielectric slab model of
laser cavity

1 per cent. and could well be expected in laser structures owing to variations in doping level, with consequent effects on the absorption coefficient and the refractive index.

Even in the earliest lasers some degree of mode confinement was, fortuitously, present. This has been enhanced, and the threshold of devices thus reduced, by devising heterojunction structures in which the composition of the semiconductor is graded and the refractive index discontinuities increased. These heterojunction devices will be considered in more detail in Chapter 6.

[iv] *Efficiency.* The electromagnetic wave in the laser cavity is partly reflected from and partly transmitted through each of the cavity mirrors. The reflected wave is amplified on each passage through the cavity and a steady state is reached in which the power lost in the cavity plus the power transmitted through the cavity mirrors is equal to the total power stimulated in the cavity. Biard, Carr and Reed[15] have analysed this situation and

deduced an expression for the external efficiency of a Fabry–Pérot cavity which gives

$$\eta_{\text{ext}} = \eta_{\text{int}} \ln\left(\frac{1}{R}\right) \left[\alpha L + \ln\left(\frac{1}{R}\right)\right]^{-1} \tag{2.45}$$

where η_{int} is the quantum efficiency of the stimulated-emission process and α is the cavity loss per unit length. It is often a good approximation to take $\eta_{\text{int}} \approx 1$.

This equation applies above threshold, and the power required to reach threshold is, essentially, wasted. For this reason the efficiency given here is termed an 'incremental' efficiency. Equation (2.45) shows that the incremental efficiency has a maximum value for small values of cavity length although, in this limit, the threshold current density of the device will rise to a high value.

[v] *Spectral properties.* The Fabry–Pérot cavity will support longitudinal modes whose wavelengths λ are given by

$$m\lambda = 2\mu L \tag{2.46}$$

where m is an integer. This represents a family of modes whose spacing $\Delta\lambda$ can be obtained by differentiating equation (2.46) with respect to m and putting $\Delta m = 1$, bearing in mind that the cavity medium is dispersive and that μ is a function of λ. Thus

$$m\frac{\mathrm{d}\lambda}{\mathrm{d}m} + \lambda = 2L\frac{\partial\mu}{\partial\lambda}\frac{\mathrm{d}\lambda}{\mathrm{d}m} \tag{2.47}$$

$$\Delta\lambda = \frac{\lambda^2}{2\mu L\left(1 - \dfrac{\lambda}{\mu}\dfrac{\partial\mu}{\partial\lambda}\right)}. \tag{2.48}$$

The wavelength of the cavity modes also varies with temperature and, by differentiating equation (2.46) with respect to temperature, the temperature coefficient can be obtained.
Thus

$$\frac{1}{\lambda}\frac{\mathrm{d}\lambda}{\mathrm{d}T} = \frac{\dfrac{1}{\mu}\dfrac{\mathrm{d}\mu}{\mathrm{d}T} + \dfrac{1}{L}\dfrac{\mathrm{d}L}{\mathrm{d}T}}{1 - \dfrac{\lambda}{\mu}\dfrac{\partial\mu}{\partial\lambda}\bigg|_T}. \tag{2.49}$$

In this expression the term involving the linear expansion coefficient of the cavity is usually negligible.

If the Fabry–Pérot cavity is considered as a three-dimensional structure, it can be seen that a mode must be defined by three integers, of which m, defining the longitudinal modes, is only one. In fact, high-resolution studies of the spectrum of a laser can reveal these additional modes. Their spacing is small compared with that of the longitudinal modes. The longitudinal modes thus appear to dominate the spectrum, and the fine structure is usually masked by imperfections in the cavity and thermal instability in the operating conditions of the device.

2.6 References

1. W. P. Dumke, *Phys. Rev.*, **132**, 1998 (1963).
2. D. G. Thomas, M. Gershenzon and F. A. Trumbore, *Phys. Rev.*, **133**, A269 (1964).
3. J. R. Haynes and N. G. Nilsson, in *Radiative Recombination in Semiconductors, Proc. Symp. Paris, 1964*, Dunod, Paris, 1965, p. 21.
4. H. Welker, *Z. Naturforsch.*, **7a**, 744, (1952).
5. J. Starkiewicz and J. W. Allen, *J. Phys. Chem. Solids*, **23**, 881 (1962).
6. N. G. Basov, O. N. Krokhin and Y. M. Popov, *Zh. Eksperim. Teor. Fiz.*, **40**, 1879 (1961) [*Sov. Phys. JETP.*, **13**, 1320 (1961)].
7. M. G. Bernard and G. Duraffourg, *Phys. Stat. Solidi*, **1**, 699 (1961).
8. G. J. Lasher and F. Stern., *Phys. Rev.* **133**, A553, (1964).
9. F. Stern, *Solid State Physics*, **15**, 364 (1963).
10. G. C. Dousmanis, C. W. Mueller and H. Nelson, *Appl. Phys. Letters*, **3**, 133 (1963).
11. G. Lucovsky, *Solid State Comm.*, **8**, 105 (1965).
12. F. Stern, *Phys. Rev.*, **148**, 186 (1966).
13. M. J. Adams, *Solid State Electron.*, **12**, 661 (1969).
14. W. W. Anderson, *J. Quantum Electron.*, **1**, 228 (1965).
15. J. R. Biard, W. N. Carr and B. S. Reed, *Trans AIME.*, **230**, 286 (1964).

2.7 Bibliography

P. J. Dean, 'A survey of radiative and non-radiative recombination mechanisms in the III–V compound semiconductors', *Trans. AIME.*, **242**, 384 (1968).

A. K. Jonscher, *Principles of Semiconductor Device Operation*, Bell, London, 1960.

T. S. Moss, *Optical Properties of Semiconductors*, Butterworth, London, 1959.

B. Ray, *II–VI Compounds*, Pergamon, Oxford, 1969.

R. A. Smith, *Semiconductors*, Cambridge University Press, 1959.

S. M. Sze, *Physics of Semiconductor Devices*, Wiley, New York, 1969.

M. Aven and J. J. Prener (Eds.), *Physics and Chemistry of II–VI Compounds*, North Holland, Amsterdam, 1967.

3

The Technology of Electroluminescent Devices

3.1 Introduction

The design and development of an electroluminescent junction device can be divided into several stages, and it will be convenient to consider these stages in general terms before showing, in a subsequent chapter, how they have been applied to particular materials and devices.

The first two stages to be considered involve the preparation of bulk single-crystal semiconductor material and the formation of an electroluminescent p–n junction in this material. These stages are essentially concerned with the production of a diode with a high internal quantum efficiency, i.e. the ratio of the number of photons produced within the junction region to the number of electrons which pass through the device should be made as high as possible. These photons will not all escape from the device, so that the third stage which is to be considered is the design of a structure that will maximize the external quantum efficiency of the device. The final stage in the procedure is the design of a device that will meet a particular requirement. Here, for example, it is necessary to distinguish between individual diodes and arrays of diodes designed to display information.

This division into four stages is somewhat arbitrary and it may well be that the stages set down cannot always be readily distinguished. Furthermore, they will almost always interact with each other, so that it is not realistic to consider them in isolation in any particular situation. However, for present purposes, it is convenient to adopt this approach.

The final section of this chapter is devoted to a discussion of the measurement of the properties of electroluminescent devices. Here there is a clear need for an understanding of the techniques used so that the experimental results which are reported can be critically compared.

3.2 The preparation of single-crystal semiconductor materials

The starting point of almost any semiconductor device, including electroluminescent devices, is a supply of the appropriate semiconductor in single-crystal form. Since the efficiency of most electroluminescent devices depends on the competition between the desired radiative processes and undesired non-radiative processes, which are favoured by impurities or crystal defects, it is clearly necessary to prepare the starting material to a high degree of purity and crystalline perfection.

The materials which are of interest all have high melting points and an appreciable vapour pressure at their melting points. These factors lead to several problems in the growth of pure single-crystal material. For example, the high temperatures involved lead to reactions between the molten semiconductor and crucible materials, giving rise to contamination of the semiconductor. Furthermore, the high pressures under which some preparations must be carried out lead to severe problems in the design of apparatus. Since, at present, most of the practical devices that have been devised are based on gallium arsenide, gallium phosphide or their ternary mixtures, most of the problems which arise and their solutions will be illustrated by reference to these materials.

Gallium arsenide has a melting point of 1240°C and exerts an arsenic partial pressure of 0·9 atmospheres at its melting point. The vapour pressure of elemental arsenic reaches the same value at 610°C and increases very rapidly with temperature. Hence, if one attempts to combine the elements gallium and arsenic by heating together the appropriate quantities to the melting point of the compound, the arsenic pressure in the system may rise to a very high value and an explosion is almost inevitable. This problem can be overcome if the elements are heated in a system containing two temperature zones, the arsenic being held at 610°C and the gallium being held at 1240°C. In this way the arsenic vapour pressure will be held at 0·9 atmospheres and the reaction with gallium will proceed to give molten gallium arsenide.

The problems encountered in making gallium phosphide are even more severe, since the melting point of gallium phosphide is 1470°C and, at this temperature, its phosphorus vapour pressure is 35 atmospheres. Elemental phosphorus reaches this vapour pressure at a temperature of approximately 550°C. Thus any system designed for the reaction of phosphorus and gallium must be able to withstand a pressure of 35 atmospheres and nowhere must the phosphorus be allowed to come in contact with a surface below 550°C.

3.2.1 The static-freeze or Stober growth technique

The static-freeze system of material synthesis and crystal growth[1] can be understood by reference to Figure 3.1. The system shown here is designed for the growth of gallium arsenide, but can obviously be used with other compounds.

A sealed quartz ampoule contains at one end a boat with a quantity of gallium and at the other end elemental arsenic. The quantity of arsenic should be sufficient to leave some excess when the reaction between it and the gallium is completed. The ampoule is placed in a multizone furnace which is designed to maintain a temperature of 610°C at one end rising to above 1240°C at the other. With this temperature profile, the reaction between arsenic and gallium is allowed to proceed until the gallium is converted to gallium arsenide. At this point the temperature of the hot zone of the furnace is progressively lowered, so that a 'freezing front' proceeds along the length of the crystal. The lower temperature of 610°C is maintained throughout this stage, so that the arsenic pressure does not drop, and a stoichiometric crystal is obtained.

The furnace must be carefully designed so that the correct temperature gradient is maintained at the freezing front. It is also important to ensure

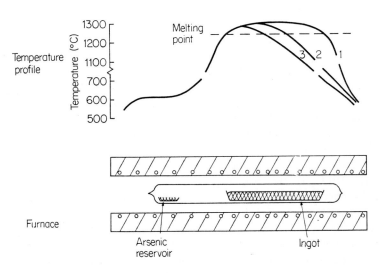

FIGURE 3.1. The static-freeze method for the preparation of gallium arsenide; profiles 1–3 represent successive conditions during the growth sequence

that this front moves continuously from one end of the boat so that solidification is not initiated at more than one point in the boat. With care, this system yields good-quality material which is largely single crystal. Doped material can be obtained by introducing the appropriate impurities into the gallium boat. In general, the dopant will be distributed along the ingot with a concentration varying monotonically from one end of the crystal to the other. This is owing to the segregation of the impurity between the solid and liquid phases and the varying impurity concentration in the liquid phase.

3.2.2 The Bridgman technique

The Bridgman technique of crystal growth,[2] which is illustrated in Figure 3.2, is similar to the static-freeze technique in the manner in which compound synthesis occurs. However, in the Bridgman technique crystal growth is made to occur by withdrawing the ampoule from the furnace. To maintain

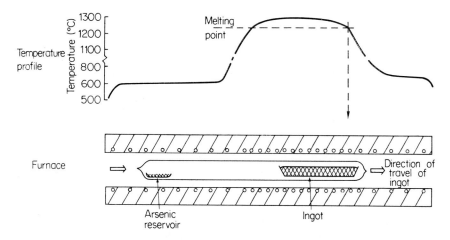

FIGURE 3.2. Bridgman technique for the preparation of gallium arsenide

the minimum temperature of the growth ampoule at the required level, the furnace must be designed to provide a flat temperature profile over a considerable length. The length of this temperature zone then sets a limit to the size of the crystal that can be produced. The furnace system must also have a very good mechanical design to ensure that the crystal is transported through the furnace without vibration. Equipment designed for the growth of gallium arsenide has been described by Cunnell, Edmond and Harding.[3]

One of the advantages of the Bridgman technique is that the temperature gradient at the freezing point remains constant throughout the crystal growth. This enables the shape of the freezing front, which is an important feature in the growth of single crystals, to be controlled. The system thus lends itself to the production of high-quality material with a low density of dislocations.

3.2.3 The Czochralski technique

The Czochralski technique[4] (Figure 3.3) is, essentially, used for the production of single-crystal material from a polycrystalline form, rather than the synthesis of a compound from its constituent elements. A quantity of the molten material is held in a crucible and into this is lowered a seed crystal of the same material. The seed is allowed to make contact with the

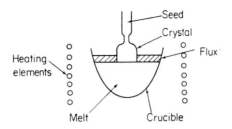

FIGURE 3.3. Czochralski technique of crystal growing (adapted for liquid-encapsulation technique)

melt and is then slowly withdrawn so that a crystal grows on the end of the seed. The grown crystal follows the orientation of the seed and, by careful control of the rate of withdrawal, coupled with a rotation of the seed, large ingots of the single-crystal material can be obtained.

The Czochralski technique is widely used in the production of silicon and other materials, but its adaption to compounds which have a high vapour pressure at their melting point raises severe problems.

For example, in a system designed to produce gallium arsenide, the whole of the inner surface of the equipment must be maintained at a temperature at or above 610°C to maintain the required arsenic pressure and thus prevent the condensation of arsenic and the decomposition of the melt. The system also has to be provided with some means of raising, lowering and rotating the seed crystal without allowing arsenic to escape. Various methods of overcoming this problem have been proposed. In one technique, the seed crystal is held in an iron chuck which can be raised and lowered by a magnet outside the system so that the system can be completely sealed. However, the most successful solution to the problem uses a technique known as liquid encapsulation,[5,6] in which the molten material is covered by a layer of a flux such as boric oxide. The diffusion of vapour through this flux is very slow, so that it is not necessary to maintain the partial pressure of the volatile element above the flux. Consequently the walls of the vessel do not have to be maintained at a high temperature, although they do have to withstand a pressure equal to the dissociation pressure of the material. The flux thus acts as an impervious membrane between the molten material, which exerts a dissociation pressure on one side, and an inert atmosphere at the same pressure on the other side.

This technique has been applied with considerable success to the growth of single-crystal gallium phosphide,[7] an example of which is shown in Figure 3.4.

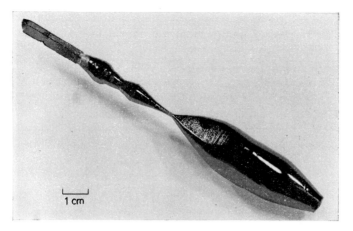

FIGURE 3.4. A crystal of gallium phosphide grown by the Czochralski technique (Crown Copyright, reproduced by permission of the Controller, Her Majesty's Stationery Office)

3.3 Epitaxial crystal-growth techniques

The development of epitaxial techniques for the preparation of semi-conductors has had a considerable influence on the development of semi-conductor devices. The basic concept of an epitaxial technique is that one commences with an orientated single-crystal slice, or substrate, of material, and grows a further layer of material on the substrate. The grown layer follows the crystal structure of the substrate—hence the term 'epitaxial', which is derived from the Greek roots epi (= upon) and taxis (= arrangement).

In epitaxial techniques, the material is grown at a temperature which is considerably below its melting point. This can reduce the contamination of the material caused by any interaction with a hot crucible and thus enables very pure material to be produced. Alternatively, trace elements can be introduced into the growth system so that doped semiconductor layers can be grown.

Epitaxial-growth processes can take place either from the liquid phase or from the gaseous phase and this forms a convenient subdivision in which to consider them in more detail.

3.3.1 Liquid-phase epitaxial growth

In the liquid-phase growth system, a saturated solution of the semi-conductor in a suitable solvent is brought into contact with an orientated substrate of the same material and the temperature of the system is slowly reduced. As the temperature falls, the solubility of the semiconductor decreases and the excess material leaves the solution and grows epitaxially on the substrate. Any impurities that are present in the melt can also be incorporated in the grown layer but, in general, the distribution of impurity between the liquid and solid phases will result in a relatively small concentration entering the epitaxial layer, which can thus be grown very pure. Conversely, to produce a doped layer, a high concentration of the trace element must be introduced into the liquid phase.

[i] *Horizontal growth.* The earliest and simplest method for carrying out a liquid-phase growth process is shown in Figure 3.5, and will be described by reference to gallium arsenide.[8]

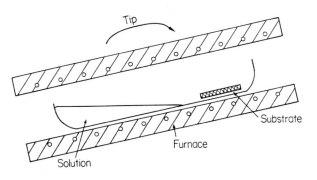

FIGURE 3.5. Liquid-phase epitaxial growth (horizontal-tipping technique)

A quantity of gallium and gallium arsenide is contained at one end of a tilted boat. A substrate slice of gallium arsenide is held at the other end of the boat. The boat is heated, either in a sealed ampoule or in a stream of hydrogen or inert gas, until the gallium arsenide is dissolved in the gallium. At a temperature of 900°C, 10 g of gallium will dissolve approximately 2 g of gallium arsenide. When the gallium is saturated, the boat is tilted so that the solution flows onto the substrate. The temperature is now allowed to fall slowly, resulting in an epitaxial growth of gallium arsenide on the substrate. When the growth has proceeded far enough the boat is tilted back to its original position, so that the remaining solution flows from the substrate and growth stops.

This somewhat crude but simple system has been considerably developed to enable thin layers of high crystal perfection to be grown. One modification to the technique is shown schematically in Figure 3.6.[9] A boat is machined from two pieces of graphite that can slide over each other. The bottom section retains the substrate in a recess, while the top section holds the melt in a

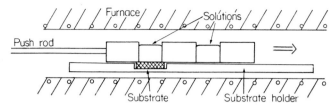

FIGURE 3.6. Liquid-phase epitaxial growth (horizontal-slide technique)

hole. After the system has been brought to the required temperature and the melt saturated, growth is initiated and stopped by sliding the melt over the substrate. Compared with the tipping method, this system gives more precise control of the growth zone and temperature, thus enabling more uniform layers to be grown. As a further development, several different melts can be contained in the same boat, so that several layers of different composition can be grown consecutively. This has been particularly valuable in the growth of laser structures.

[ii] *Vertical growth*. In another version of the liquid-phase growth technique,[10] a substrate is lowered into the melt (Figure 3.7). The substrate

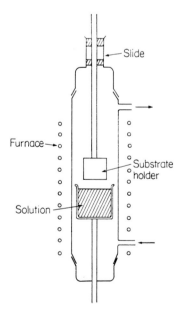

FIGURE 3.7. Liquid-phase epitaxial growth (vertical technique)

is held against a quartz plate which is on the end of a rod. This rod is a sliding fit in a seal at the top of the apparatus, so that the seed can be raised or lowered as required.

3.3.2 *Vapour-phase epitaxial growth*

Several techniques which enable semiconductor materials to be grown from the vapour phase have been developed. In some cases a convenient starting point is a polycrystalline or powder form of the semiconductor compound. This is transported by a vapour-phase reaction to a nucleating surface or a suitable substrate on which a single-crystal layer of the compound can grow. More commonly, one starts with readily purifiable compounds of the elements that comprise the semiconducting material. These are made to undergo chemical reactions to produce the semiconductor and deposit it as a single-crystal layer on a suitable substrate. Two such processes have found particular favour in work on the III–V compounds.

[i] *The halide system.*[11] The chlorides of phosphorus and arsenic, PCl_3 and $AsCl_3$, can be obtained in a highly purified form and are thus convenient starting materials for the synthesis of GaP and GaAs. A system for the epitaxial growth of GaAs is illustrated in Figure 3.8.

Hydrogen, which is purified by diffusion through palladium, is bubbled through $AsCl_3$ in a controlled-temperature bath. It is then passed over a boat containing gallium at 800°C. The initial reaction in this part of the furnace saturates the gallium with arsenic, while waste chlorine compounds pass through the furnace. The gallium becomes saturated with arsenic at a concentration of a few atomic per cent. and, as soon as this stage is reached, gallium arsenide is transported along the tube. If a seed crystal is placed in

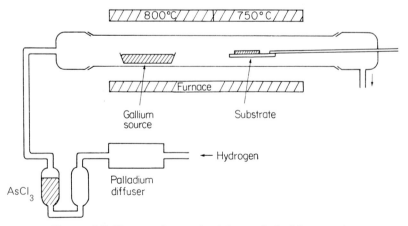

FIGURE 3.8. Vapour-phase epitaxial growth (halide system)

the vapour stream in a cooler region of the furnace, gallium arsenide will be deposited from the vapour phase and grow as an epitaxial layer on the seed.

The material which is grown can be doped by introducing donor or acceptor elements either into the gallium source or as volatile compounds into the gas stream.

The halide system can also be used to grow gallium phosphide using PCl_3 in place of the $AsCl_3$, and gallium arsenide–phosphide alloys can be grown by passing parallel gas streams through $AsCl_3$ and PCl_3. Here it is important to control accurately the temperature of the halides, as their vapour pressures depend exponentially on temperature, making it somewhat difficult to control the composition of the grown material.

[ii] *The hydride system.*[12] A schematic diagram of equipment designed to synthesize and grow gallium arsenide or gallium phosphide from the hydrides arsine (AsH_3) or phosphine (PH_3) and gallium is shown in Figure 3.9.

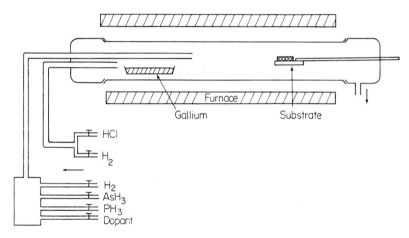

FIGURE 3.9. Vapour-phase epitaxial growth (hydride system for GaAs or GaAsP)

For gallium arsenide, a stream of HCl passes over a gallium source at 800°C and transports gallium as a volatile chloride along the furnace tube. A stream of arsine bypasses the gallium boat and decomposes in the hot zone of the furnace, where the two vapour streams mix. GaAs is formed in a cooler part of the tube and is deposited on a suitable substrate.

As in the halide growth system, doped material can be grown by introducing gaseous forms of the appropriate trace elements. Gallium phosphide can be grown by using PH_3 in place of AsH_3 and, by using appropriate mixtures of PH_3 and AsH_3, mixed-alloy compounds can be grown. In this respect the method is particularly convenient, as the control of the relative quantities of AsH_3 and PH_3 enables the composition of the grown material to be controlled and varied at will. The composition can even be changed during growth, as the gallium source does not have to establish equilibrium with the arsenic or phosphorus vapour. One of the disadvantages of the system is the difficulty of obtaining gaseous HCl and the hydrides of the required purity.

3.4   The formation of *p–n* junctions

A semiconductor *p–n* junction is the boundary region between *p*-type and *n*-type materials. The formation of a *p–n* junction is thus dependent on the ability to make the semiconductor both *p* type and *n* type, and this is achieved by the introduction of suitable trace elements or dopants into the semi-conductor.

3.4.1   *Donor and acceptor elements*

The selection of suitable trace elements to dope a semiconductor can be understood by reference to the periodic table, a portion of which is given in Table 2.2. In general, to dope a semiconductor some of the atoms of the semiconductor host lattice must be replaced by suitable impurity atoms. If a dopant atom has more valence electrons than the atom it replaces, it will be able to donate one or more electrons to the conduction band of the semiconductor. Such an element, known as a donor, will make the material *n* type.

For example, in germanium, which is in group IV of the periodic table, the group V element phosphorus is a donor. In the III–V compounds, the group VI elements sulphur, selenium and tellurium will enter the crystal lattice on the group V element sites and are donors. Similarly, if silicon replaces the group III element in a III–V material, it will act as a donor.

Conversely, an atom with fewer valence electrons than the element it replaces will act as an acceptor and produce *p*-type material. For the III–V compounds, the most commonly encountered acceptors are the group II elements zinc and cadmium. Silicon, if it replaces the group V element, will also behave as an acceptor.

In these examples, silicon is an amphoteric dopant in gallium arsenide, i.e. it can act as either a donor or an acceptor, depending on the lattice site that it occupies. It is possible to control the way in which such an impurity enters the lattice and it will be seen that this is of considerable importance in some devices made from gallium arsenide.

In the II–VI compounds the situation is not so simple. In principle, similar considerations apply to the selection of donor and acceptor impurities, but it is found that many of the II–VI compounds have an inherent tendency to remain of one conductivity type, irrespective of the dopant elements that are introduced. This arises because the electrical properties of the added elements are compensated by the electrical properties of lattice defects that are also formed.

3.4.2 Epitaxial junctions

In principle, the simplest techniques available for the formation of a *p–n* junction are based on epitaxial processes. It has been explained that donor or acceptor elements can be added to the melt in a liquid-phase system or introduced into the gas stream in a vapour-phase system, and thus thin layers of the required conductivity type can be grown. In this way, a layer of *p*-type material can be grown on an *n*-type substrate (or vice versa) to give a *p–n* junction [Figure 3.10(a)].

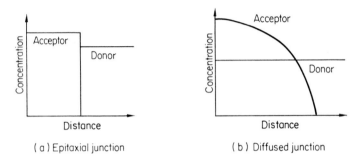

(a) Epitaxial junction   (b) Diffused junction

FIGURE 3.10. Doping profiles in *p–n* junctions

This technique leads to one problem, in that the *p–n* junction occurs in a region which formed the original surface of the material. Unless extreme care is taken in the preparation of this surface, the junction will be formed in a region of poor crystal structure, since the surface of a crystal is rarely free from damage.

The problem can be minimized by using a technique, known as 'etchback', in which a small amount of the substrate surface is removed before growth commences. In a liquid-phase system, this can be achieved by allowing an unsaturated melt to meet the substrate which will thus dissolve some of it, while, in a vapour-phase system, it can be achieved by initially raising the temperature of the seed. Alternatively, the problem can be eliminated by changing the dopant element during growth, so that the complete *p–n* structure is grown. This can easily be achieved by using a vapour-phase growth system in which the dopants are introduced into the gas flow and can readily be controlled.

Another problem which can arise when junctions are grown by epitaxial techniques lies in the accurate control of the doping level. The concentration of an impurity that is incorporated in a crystal grown from liquid or vapour phases depends on several factors, such as the concentration of impurity in the fluid phase, the growth temperature and temperature gradients in the system. In this respect, the liquid-phase technique is particularly difficult to control, as the growth conditions are changing throughout the growth process. The vapour-phase technique is easier to control, as the epitaxial process proceeds under constant-growth conditions.

One of the big advantages of epitaxial junction-formation techniques is that they allow complete freedom of choice of the doping levels on either side of the junctions. This is not the case with the diffusion techniques that are considered in the next section.

3.4.3 *Diffused junctions*

The diffusion technique is probably the most widely used and convenient method available for the formation of *p–n* junctions. In essence, one starts with material doped to one conductivity type, and then introduces impurities to overdope the material and produce the opposite conductivity type. This is achieved by diffusing impurities from the surface of the material [Figure 3.10(b)]. With a semiconductor such as silicon, which is stable at high temperatures, the process can be very simple; it is merely necessary to heat the material in an atmosphere containing the required dopant. The situation with the compound semiconductors is more complicated because, at the temperature required for the diffusion process, the semiconductor will dissociate to an appreciable extent. To avoid this, it is necessary to perform the diffusion process in an atmosphere containing not only the required diffusant, but also the volatile element of the compound so as to suppress dissociation. The process is most conveniently carried out by enclosing the semiconductor in an ampoule containing suitable amounts of the appropriate elements. Alternatively, it is possible to coat the semiconductor with a layer of the diffusant in a solid form. This layer can then inhibit the dissociation of the material underneath it and, at the same time, provide a source for the diffusion process.

During the initial stages of a diffusion process, the diffusant establishes equilibrium between the vapour phase and the semiconductor surface. The subsequent diffusion of material from this surface into the bulk of the semiconductor can be described by Fick's law, which relates the rate of transport of material to the concentration gradient of that material and the diffusion coefficient *D*. For the one-dimensional case, Fick's law can be written as:

$$\frac{\partial C}{\partial t} = D \frac{\partial^2 C}{\partial x^2} . \tag{3.1}$$

The solution of this equation gives the variation of the concentration *C* as a function of distance below the surface of the semiconductor *x* and the diffusion time t. For the simple case, in which the surface concentration C_0 remains constant and the diffusion coefficient D is independent of concentration, this concentration is given by the complementary error function

$$C = C_0 \text{ erfc } [x/2\sqrt{(Dt)}]. \qquad (3.2)$$

In practice, this simple situation rarely exists and, in most cases, equation (3.2) does not accurately describe the observed diffusion profile. Extensive studies of the diffusion of impurities in the III–V compounds have been reported and this work has been reviewed by Kendall.[13]

A typical example is the diffusion of zinc into gallium arsenide. This has been extensively studied, and some experimental results are given in Figure 3.11. The series of diffusion profiles shown here was obtained by varying the zinc vapour pressure surrounding the gallium arsenide while keeping the diffusion time and temperature constant.[14] It can be seen that the zinc concentration falls slowly in the region close to the surface, but drops more rapidly after a concentration of $3 \times 10^{19} \text{cm}^{-3}$ has been reached.

The form of these profiles can be explained by postulating that zinc can occupy either substitutional or interstitial sites in the gallium arsenide lattice. In equilibrium, the concentration of substitutional species is much greater than that of the interstitial species, but the diffusion process is dominated by the transport of the more mobile interstitial atoms. A detailed

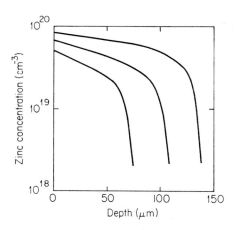

FIGURE 3.11. Diffusion profiles in gallium arsenide (zinc diffused from vapour phase, diffusion temperature = 1000°C, diffusion time = 3 h); the parameter varied is the zinc vapour pressure

analysis of this model predicts an effective diffusion coefficient that is dependent on the zinc concentration and can thus explain the shape of the diffusion profiles.

The arsenic pressure in the ampoule surrounding the crystal also affects the diffusion process. A high arsenic pressure will increase the equilibrium concentration of gallium vacancies in the lattice, and more of the zinc atoms will occupy these sites. Thus the net rate of diffusion, which depends on the number of interstitial atoms, is reduced.

Although the diffusion processes that occur in other compound semi-conductors have not been studied in as much detail as this example, similar considerations are believed to apply in many cases. However, since a detailed analysis of any particular situation is such a complex problem, the diffusion conditions necessary to achieve any particular concentration profile have usually been derived on an empirical basis.

In principle, p–n junctions could be formed either by the diffusion of acceptors into n-type material or by the diffusion of donors into p-type material. However, with the III–V compounds, the diffusion of metallic acceptor impurities such as zinc or cadmium proceeds at a convenient rate, whereas the diffusion of the group VI donor elements is inconveniently slow. Thus the diffusion of the acceptor impurities has been more extensively studied and is more widely used in device technology.

3.4.4 Planar techniques

Planar techniques, which have found wide application in the technology of silicon devices, have also been used to fabricate some electroluminescent devices. The basic idea behind this technique is to deposit, on the surface of the semiconductor, a material that will inhibit the diffusion of impurities into the semiconductor. This impervious layer can be restricted to particular areas of the semiconductor, so that diffusion can take place only at the unprotected surface. In this way, isolated diffused junctions can be produced. An outline of the basis of the planar technology is shown in Figure 3.12.

The first stage is to deposit an impervious diffusion mask over the semi-conductor surface, and for this purpose layers of silica or silicon nitride can be used. In the technology of silicon devices, suitable layers can be produced by thermally oxidizing the surface of the silicon, or, alternatively, the mask can be deposited onto the surface. With other semiconductors only deposition techniques are applicable. To act as a diffusion mask a layer 5000 Å thick is required, and suitable layers can be produced by a number of techniques:

(a) By evaporation or sputtering.[15]
(b) By the decomposition of silane, in the presence of oxygen or ammonia, to give, respectively, SiO_2 or Si_3N_4 layers. The decomposition can be induced thermally[16] or by an r.f. discharge.[17]

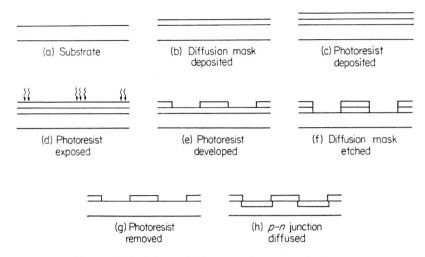

(a) Substrate

(b) Diffusion mask
deposited

(c) Photoresist
deposited

(d) Photoresist
exposed

(e) Photoresist
developed

(f) Diffusion mask
etched

(g) Photoresist
removed

(h) *p–n* junction
diffused

FIGURE 3.12. Schematic diagram of planar technology

(c) By the thermal decomposition of tetraethoxysilane (TEOS) to give silica
layers[18] (Figure 3.13).

After the diffusion mask has been deposited, holes or windows must be
etched in it to allow diffusion to take place over the required areas. These
holes can be cut by photolithographic techniques, in which a photosensitive
layer (photoresist) is deposited on the mask. This is then exposed photo-
graphically and the unexposed regions removed, thus allowing a subsequent
etch to remove the underlying diffusion mask. The exposed regions of the
photoresist are removed, and the diffusion process carried out.

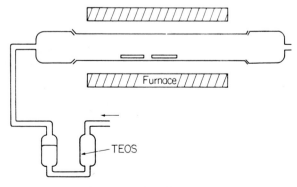

FIGURE 3.13. Apparatus for the deposition of SiO_2
from TEOS

Since photographic techniques allow complex patterns of diffusion windows to be made, planar technology is particularly suitable for making arrays of a number of diodes in one substrate. The exploitation of this aspect of the technology will be considered in more detail in a later section.

An important feature of the planar technique is the protection that it affords the p–n junction. It is well known that the region where a p–n junction meets the atmosphere is particularly susceptible to contamination which can degrade the performance of the device. With a planar diffused junction this boundary region occurs under the surface oxide layer, which thus gives protection to the edge of the junction. Diodes made in this way have been shown to be extremely reliable and free from degradation.

3.4.5 *Ion implantation*

In recent years, the technique of ion implantation has been developed as a means of producing complex doping profiles in semiconductor devices. A beam of ions of the required dopant element is accelerated to a potential of 10–250 keV and made to strike the surface of the semiconductor. The ions travel a finite distance, of the order of a few micrometres, before coming to rest. The final distribution of dopant atoms, which depends on the nature and energy of the ion and the orientation of the semiconductor, can be predicted accurately. Thus, by varying the ion-accelerating voltage during the implantation, the penetration depth can be varied and complex doping profiles can be produced.

The implantation process damages the crystal lattice to a considerable extent, but this damage can be removed by annealing the crystal. The temperature required for the annealing process is usually considerably lower than that required for a diffusion process, so that the ion-implantation technique can be used to produce p–n junctions in situations where the use of a high temperature is unacceptable.

At present, no practical electroluminescent devices have been based on ion implantation, but the technique is being explored as a possible way of making p–n junctions in the II–VI compounds. p–n junctions cannot be made by diffusion or epitaxial deposition of many interesting II–VI materials because an autocompensation mechanism restricts the conductivity type of the material that can be obtained. However, there is some evidence that the autocompensation mechanism is ineffective at the lower temperatures required for ion implantation. Thus it may be possible to produce electroluminescent p–n junctions in materials such as ZnTe.

3.5 Optical, electrical and thermal problems in the design of electroluminescent diodes

Although it may be possible to make p–n junctions in which the electroluminescent efficiency is high, even approaching 100 per cent. in some cases, the problem of extracting radiation from the diode and thus making an efficient device requires further consideration. Often the requirements for making an efficient device conflict with those for a useful or practical structure, and it must be borne in mind that economic considerations also play an important part in the design of a device.

In this section consideration will be given to some of the problems that are encountered in trying to meet these conflicting requirements.

3.5.1 Optical absorption

Any radiation produced in the p–n junction region of an electroluminescent device must travel through a finite thickness of semiconductor before being emitted from the device. The radiation will thus suffer some absorption, and it is important to keep the path of radiation within the semiconductor as small as possible. It is difficult to determine the value of the absorption coefficient that is applicable to any particular situation, since this depends, in a complex way, on several factors. However, some of the relevant factors can be discussed in qualitative terms, and these, together with published experimental results, give adequate guidance for the design of device structures.

The optical absorption coefficient of a semiconductor depends strongly on the wavelength of the radiation, particularly in the spectral region which corresponds to the band gap of the material. In a typical example, the absorption coefficient can rise from about 10 cm^{-1} on the long-wavelength side of the band edge to 10^3cm^{-1} at shorter wavelengths. The radiation in this spectral region will be attenuated by a factor of $1/e$ in a path length of between 10 and 1000 μm.

The absorption coefficient of a semiconductor also depends on the impurity concentration within the material. Ionized donors or acceptors will populate the edges of the conduction or valence bands with electrons or holes, and these occupied band states will not be able to take part in an absorption process. The result of this is to shift the apparent absorption edge to a higher energy, an effect which is known as the Moss–Burstein shift.

At high doping levels a further complication occurs. As the donor or acceptor concentration is increased, the impurity levels broaden until they interact with the edges of the energy bands, with a consequent effect on the absorption coefficient.

The rapid variation of absorption coefficient with photon energy in the vicinity of the band edge can be used to minimize the absorption of radiation in devices that are made of ternary alloy semiconductors. In these devices, it may be possible to vary the composition of the alloy in the region adjacent to the p–n junction. If this material is graded to a composition that has a wider gap, the optical absorption as radiation traverses this region can be considerably reduced.

In a diffused diode, the impurity concentration will usually be much higher on one side of the junction than the other and this will determine which side of the diode should be used as the 'window' for the radiation. For example, in a gallium arsenide diode that is formed by the diffusion of the acceptor zinc into n-type material, the heavily doped p-type layer has a much higher absorption coefficient, at the radiation wavelength, than the n-type material. Thus the device should be designed, if possible, so that the n-type material is used as the device window. If this is not possible, the p-type region must be kept thin by a careful choice of the diffusion conditions.

3.5.2   Refractive index mismatch

Both reflection and refraction occur when radiation meets a discontinuity in refractive index and, since semiconductors have a high refractive index, these effects will be pronounced at the surface of an electroluminescent diode. The refractive index data of some semiconductors are given in Table 3.1.

TABLE 3.1. The refractive indices of some semiconductors (values are at room temperature and for wavelengths near the absorption edge of the material).

GaP	3·4	ZnS	2·3	CdS	2·3
GaAs	3·6	ZnSe	2·4	CdSe	2·5
InP	3·4	ZnTe	2·7	CdTe	2·7

For the III–V compounds, the refractive index n is approximately 3·5 at wavelengths close to the absorption edges of the materials, and the critical angle for total internal reflection in these materials is approximately 15°. Further data for the III–V compounds have been compiled by Seraphin and Bennett.[19]

Only radiation that meets the surface of a device at an angle less than the critical angle will emerge, and thus a large fraction of the radiation will be reflected back into the semiconductor. This will result in a considerable loss in the efficiency of the device. The problem can be minimized by carefully designing the shape of the device so that as much of the radiation as possible meets the semiconductor surface at small angles. However, it is often not only the efficiency of a device that is important, but also the intensity of emission in a particular direction and the apparent luminance of the source. The shape of the device surface can affect both of these properties. Frequently a structure designed to increase the amount of radiation emitted by a device also produces a magnified image of the source, and thus the luminance may be unchanged, even though the efficiency is increased. The geometrical properties of three commonly used structures will be discussed; further details of these and other structures can be found in Reference 20.

[i] *Plane-slab structure.* One of the simplest device structures is that shown in Figure 3.14. Radiation is emitted isotropically from all points on a junction plane that lies within a slab of semiconductor. It will be assumed that all the surfaces of this structure are flat and optically perfect.

If a unit area of the junction emits a power P, the power which meets one surface at an angle less than the critical angle C is given by

$$PF = \frac{P\int_0^C \sin\theta d\theta}{2\int_0^{\pi/2}\sin\theta d\theta} = \frac{P(1-\cos C)}{2} \approx \frac{P}{4n^2}. \tag{3.3}$$

The transmission, at normal incidence, of a dielectric–air interface is given by

$$T = 1 - \left(\frac{n-1}{n+1}\right)^2 = \frac{4n}{(n+1)^2}. \tag{3.4}$$

The value of this transmission coefficient changes little for small angles up to the critical angle. Hence the power that emerges from the device is given by

$$P_0 = PFT$$

$$= \frac{P(1-\cos C)}{2} \frac{4n}{(n+1)^2} \approx \frac{P}{n(n+1)^2}. \tag{3.5}$$

Taking gallium phosphide as an example, only 3 per cent. of the radiation is contained within the subcritical cone, and only 72 per cent. of this is transmitted through the semiconductor surface. If the radiation suffers any appreciable absorption in the semiconductor, the efficiency will be even lower.

In some situations, it may be possible to utilize the radiation that is emitted from all the surfaces of the device, and then the device efficiency is increased by a factor of six. In other situations, the lower surface of the

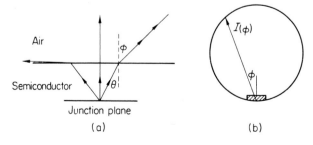

FIGURE 3.14. The simple slab device structure, showing (a) ray paths in the structure and (b) a polar diagram of the radiation intensity

slab may be made a perfect reflector, so that radiation is reflected back to the top surface. This will increase the device efficiency by a factor of two.

The variation of the intensity of emission with direction from the simple slab structure can be calculated as follows. Suppose radiation is emitted from all points on the junction with an intensity distribution $I(\theta)$ and emerges from the device with a distribution $I(\phi)$. The radiation will meet the device surface at an angle θ and emerge at the refracted angle ϕ. Considering an elementary cone of radiation in solid angle $d\theta \times d\theta$, this will emerge as a cone of solid angle $d\theta \times d.\theta$ By the laws of refraction

$$\sin \theta = n \sin \theta$$

$$\cos \theta \, d\theta = n \cos \theta \, d\theta.$$

Also, since the total flux contained in the cone is constant,

$$I(\theta) \, d\theta \, d\theta = I(\theta) \, d\theta \, d\theta.$$

Thus
$$I(\theta) = \frac{I(\theta) \cos \theta}{n \cos \theta}. \tag{3.6}$$

For large values of n, $\cos \theta$ is close to one up to the critical angle, and, for an isotropic emitter, $I(\theta)$ is independent of θ.
Thus

$$I(\theta) = I_0 \cos\theta \tag{3.7}$$

where I_0 [$= I(\theta)/n$] is the intensity of radiation normal to the device surface. The source thus obeys Lambert's cosine law, and has a luminance that is independent of the direction θ. As for any Lambertian source, a simple integration shows that the total power emitted is given by

$$P_0 = \pi I_0.$$

Thus, from equation (3.5)

$$I_0 = \frac{1}{\pi} P_0$$

$$= \frac{P(1 - \cos C) \, 4n}{2\pi(n+1)^2}$$

$$\approx \frac{P}{\pi n(n+1)^2}. \tag{3.8}$$

[ii] *Hemispherical structure.* Considerable thought has been given to the problem of shaping the surface of a diode to minimize the problems caused by total internal reflection. If, for example, the diode were placed at the centre of a large sphere of the same material, all the radiation would meet the surface at normal incidence, and would thus not be subject to total internal reflection. If this sphere were made too large, radiation would be

absorbed in the device, while, if the sphere were too small, some radiation would still suffer total internal reflection.

It can be shown that, for a junction of diameter d, emitting radiation at the centre of a sphere of diameter D, the maximum angle of incidence at the surface which can occur is given by $\tan^{-1} d/D$. Hence, for no internal reflection to occur, the sphere must have a diameter $D = nd$. In this situation, radiation that is emitted isotropically from the junction will emerge from the device with an approximately isotropic radiation pattern (Figure 3.15). The power that is emitted from the hemisphere is given by

$$P_0 = \frac{PT}{2} = \frac{2nP}{(n+1)^2} \tag{3.9}$$

and is increased by a factor of $2n^2$ over that from the plane device considered in the previous section.

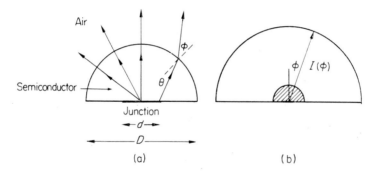

FIGURE 3.15. The hemispherical device structure, showing (a) ray paths in the structure and (b) a polar diagram of the radiation intensity

The intensity of radiation, which is isotropic, is thus given by

$$I_0 = \frac{P_0}{2\pi} = \frac{PT}{4} = \frac{Pn}{\pi(n+1)^2} \tag{3.10}$$

and is increased by a factor of n^2 over that from a plane device. However the hemispherical semiconductor lens increases the apparent size of the source by a linear factor of n; so that the luminance of the hemispherical device in the direction perpendicular to the junction plane is the same as that of the plane device.

[iii] *The Weierstrass sphere.* An interesting device structure,[21] which utilizes a spherical geometry, and is known as the Weierstrass sphere, is shown in

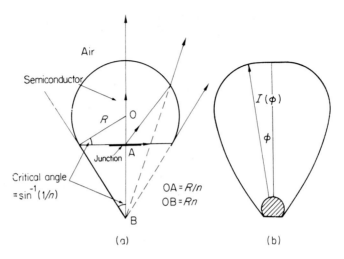

FIGURE 3.16. The Weierstrass-sphere device structure, showing
(a) ray paths in the structure and (b) a polar diagram of the
radiation intensity

Figure 3.16. In this structure, two aplanatic points A and B are situated at
distances R/n and nR from the centre of a spherical surface of radius R.
These are conjugate points, and any ray originating at one point appears,
externally, to have originated at the other point. Furthermore, for a
sufficiently large sphere, all the radiation from A will meet the spherical
surface at less than the critical angle and will not be totally internally
reflected. In a device structure, the radiating junction plane passes through A.

From Figure 3.16, it can be seen that radiation emitted from a junction
point into a hemisphere emerges from the device within a cone of semiangle
$\sin^{-1}(1/n)$ radians. The solid angle in this cone is $2\pi[1-\cos(1/n)] \approx \pi/n^2$
steradians.

The power emitted by the device is, to a good approximation, given by

$$P_0 = \frac{PT}{2} = \frac{2nP}{(n+1)^2}.$$

(3.11)

Thus the mean intensity of radiation in the solid angle π/n^2 is given by

$$I_0 = \frac{PTn^2}{2\pi} = \frac{2Pn^3}{(n+1)^2}$$

(3.12)

and is increased by a factor of $2n^2$ over that from a simple hemispherical
structure.

The Weierstrass sphere magnifies the radiation source by a linear factor of n^2, so that the mean luminance of the source is a factor of two greater than that of both the plane and the spherical sources.

The properties of the structures that have been described in the previous sections are summarized in Table 3.2. Although the formulae that have been derived involve several approximations, they are sufficiently accurate for most purposes, and are almost certainly as accurate as any relevant experimental results.

The effect of multiple reflections has also not been considered, and it has been assumed that any radiation that does not emerge from the device on its first encounter with a surface will not emerge subsequently. This may be true in some situations, but, alternatively, radiation may traverse the structure several times, enhancing the probability that it will emerge from the device.

In practice it is difficult, and therefore expensive, to shape the surface of a semiconductor device to the required profile, and the usual approach that is adopted is to encapsulate the diode in a transparent resin of high refractive index. This resin can be moulded to the desired shape. The transparent encapsulating materials that are readily available have refractive indices of, 1·5–1·6, and, although this falls far short of the values for semiconductors, encapsulation with such a resin increases the output by a factor of two to three. Attempts have been made to find transparent materials with higher refractive indices that could be used as encapsulants. For example,[22] glasses

TABLE 3.2. Figures of merit for device structures. Intensity of radiation and luminance values are calculated for unit power generation per unit junction area. Numerical values are for a refractive index $n = 3\cdot6$.

Source geometry	Efficiency	Intensity of radiation	Magnification of source area	Luminance (on axis)
Plane	$\dfrac{1}{n(n+1)^2}$ $(= 0\cdot013)$	$\dfrac{1}{\pi n(n+1)^2}$ $(= 0\cdot0042)$	1 $(= 1)$	$\dfrac{1}{\pi n(n+1)^2}$ $(= 0\cdot0042)$
Hemisphere	$\dfrac{2n}{(n+1)^2}$ $(= 0\cdot34)$	$\dfrac{n}{\pi(n+1)^2}$ $(= 0\cdot055)$	n^2 $(= 13)$	$\dfrac{1}{\pi n(n+1)^2}$ $(= 0\cdot0042)$
Weierstrass sphere	$\dfrac{2n}{(n+1)^2}$ $(= 0\cdot34)$	$\dfrac{n^3}{\pi(n+1)^2}$ $(= 1\cdot4)$	n^4 $(= 170)$	$\dfrac{1}{\pi n(n+1)^2}$ $(= 0\cdot0084)$

with As–S–Br or As–S–Se–Br compositions can have refractive indices of about 2·5, but these usually give a poor thermal-expansion match with the semiconductor and are difficult to use.

3.5.3   Electical contacts

An electroluminescent diode must be provided with two electrical contacts. The contacts must be designed to keep the series resistance of the device to a low value but not obstruct the radiation emerging from the junction. These two requirements usually impose conflicting criteria on the structure.

One of the most convenient device structures is shown in Figure 3.17. In this structure, the contact to the *n*-type material covers one side of the

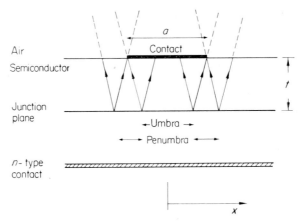

FIGURE 3.17. The obstruction of radiation by device contacts

semiconductor die, and this surface is usually bonded to a heat sink. The other contact, to the *p*-type material, covers a small area of the top surface. This contact may take one of a variety of forms, ranging from a narrow stripe in the centre or around the edge of the die to a complex grid pattern across the surface.

The conflicting requirements of this structure can easily be identified. To obtain a low resistance, the area of the contact should be made as large as possible, but, on the other hand, to obstruct the minimum amount of radiation the contact needs to be small. To formulate a simple quantitative example of this problem, it will be assumed that the current through the device is uniformly distributed across the junction, although this will often be a poor approximation to the real situation.

As has been seen in Section 3.5.2, radiation can only emerge from the device if it leaves the junction within a cone whose semiangle C is the critical angle for the semiconductor–air interface. Radiation outside this cone will be totally reflected at the semiconductor surface. With the dimensions shown in Figure 3.17, radiation can emerge unobstructed from the device if it is produced within the junction region defined by $|x| > \frac{1}{2}a + t\tan C$ and is completely obscured from the umbra region defined by $|x| < \frac{1}{2}a - t\tan C$. Radiation from the penumbra regions is partly obscured.

In practice, the current through the *p–n* junction will not be distributed uniformly, but will be 'crowded' under the contact to a greater or lesser extent, depending on the resistivity and thickness of the semiconductor layer. The crowding will be least when the resistivity of the layer is low and the layer is thick, but both of these factors will increase the optical absorption in the structure.

For a contact of a given area, this situation can be improved by subdividing the contacts into a number of narrow regions. Although this approach does not affect the total area of the junction that is obscured, it will improve the uniformity of the current distribution and thus increase the light output of the device. The logical extension of this approach is to make the contact in the form of a fine grid extending over the surface of the device. However, in most practical devices, this is not necessary, and an adequate structure can be obtained with a relatively simple contact pattern.

The electrical contacts are usually made by depositing metallic layers onto the semiconductor, which is then heated to form an alloy between the semiconductor and the metal. The basic metals most often used for the contacts are gold or silver, which form alloys with the III–V compounds at temperatures around 500°C. Small proportions of other metals are often added to the basic contact metal to improve the metallurgical and electrical properties of the contact.

The most convenient way of depositing the contact layers is by means of evaporation. Either, all the metals can be evaporated simultaneously from an alloy source, or the required material can be evaporated as successive layers. To form contacts with a complex pattern, a mask can be used in the evaporation process, or the contact can be delineated using conventional photolithographic techniques.

3.5.4 *Thermal properties of devices*

Electroluminescent diodes operate at current densities in the range 1–10 A/cm^2, and the corresponding power densities are approximately 2–20 W/cm^2. These power densities are sufficient to cause an appreciable temperature rise in the device and, since diode efficiencies, in general, decrease rapidly with increasing temperature, it is important to design device structures that will minimize this temperature rise. This is even more important for lasers, which may operate at current densities as great as 10^4 A/cm^2.

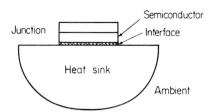

FIGURE 3.18. Thermal structure of a
diode

[i] *Thermal structure of diodes.* The thermal impedance of the diode shown schematically in Figure 3.18 can be considered as several components in series.

[a] The semiconductor. The thermal conductivity of most semiconductors at room temperature lies in the range $0 \cdot 1$–1 W/cm K. The values for some materials that are of interest are given in Table 3.3, and further data for the III–V compounds have been compiled by Holland.[23] It is important to note that the thermal conductivity of the ternary III–V materials is considerably lower than that of either binary constituent. For example, in the case of $GaAs_x P_{1-x}$, shown in Figure 3.19, the material of composition $GaAs_{0.5} P_{0.5}$ has a thermal conductivity which is nearly an order of magnitude lower than that of either GaAs or GaP.

TABLE 3.3. The thermal conductivities and thermal expansion coefficients of some
semiconductors and metals at room temperature.

	Materials	Thermal conductivity (W/cm K)	Linear expansion coefficient (K^{-1})
Semiconductors	Si	1·41	$2 \cdot 3 \times 10^{-6}$
	Ge	0·61	$5 \cdot 7 \times 10^{-6}$
	GaP	0·77	5×10^{-6}
	GaAs	0·46	6×10^{-6}
	ZnSe	0·19	7×10^{-6}
	CdTe	0·18	5×10^{-6}
Metals	Ag	4·2	$1 \cdot 9 \times 10^{-5}$
	Al	2·0	$2 \cdot 3 \times 10^{-5}$
	Au	3·0	$1 \cdot 4 \times 10^{-5}$
	Cu	3·9	$1 \cdot 7 \times 10^{-5}$
	Mo	1·5	$4 \cdot 9 \times 10^{-6}$
	W	1·8	$4 \cdot 3 \times 10^{-6}$

When designing a diode, it is usually possible to keep the thermal impedance of the semiconductor region small by using only thin layers of material and restricting the use of mixed-alloy components to the active junction region. For example, thin layers of $GaAs_xP_{1-x}$ can be grown on GaAs substrates.

[b] The semiconductor–heat-sink interface. The interface between the semiconductor diode and the heat sink usually comprises layers of metallic contacts and solder. If the contact is badly made, so that, for example, the solder only wets restricted areas of the device, a poor thermal structure will result. However, in general, this region does not present a serious thermal barrier.

[c] The heat sink. When a heat flux flows through a small area into a semi-infinite solid, a finite thermal spreading resistance occurs. This resistance depends not only on the area of the contact, but also on the shape

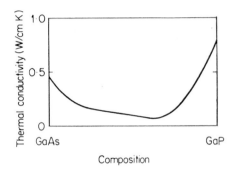

FIGURE 3.19. Thermal conductivity of
GaAs–GaP at 300 K

of the contact.[24] For example, a rectangular contact of dimensions $l \times w$ (where $l > w$) on a heat sink of thermal conductivity K will have a thermal resistance given by

$$\Theta = \frac{1}{\sqrt{2\pi l K}}\ln(4l/w). \tag{3.13}$$

This expression shows that, for a given area, a long thin contact will have less resistance than a square contact. However, the device designer rarely has this freedom of choice, and most devices have a square or round junction area.

For a circular contact of radius a, the thermal impedance is given by

$$\Theta = \frac{1}{4Ka} \tag{3.14}$$

showing that, in this configuration, the thermal impedance is inversely proportional to the linear dimension of the contact.

The heat-sink material cannot always be chosen with regard only to its thermal conductivity. It is usually necessary to choose a material that has a thermal-expansion coefficient close to that of the semiconductor. This usually rules out good thermal conductors such as copper or aluminium, whose expansion coefficients are much larger than those of the semiconductors (Table 3.3), and necessitates the use of poorer conductors, such as molybdenum.

In many situations, the major contribution to the thermal impedance of a device lies in the interface between the heat sink and the ambient atmosphere. To minimize this, devices can be attached to heat sinks of large surface area such as are used for any semiconductor power device.

[ii] *The measurement of thermal impedance.* In most semiconductor diodes, the forward bias V' at a given forward current I' is a linear function of the junction temperature. A measurement of the voltage V' can thus be used to monitor the temperature of the p–n junction. To measure the thermal impedance of a device, it is first necessary to measure V' as a function of temperature T. The device is then operated at its design current I, which is reduced to the value I' for a short time, during which V' is measured. Provided that the low-current pulse is short enough not to disturb the junction temperature, this temperature, and hence the device thermal impedance, can be determined.

[iii] *Power dissipation.* The current–voltage characteristic of a diode can be considered approximately as a p–n junction operating at a constant bias V_0 in series with a resistance R. The dissipation of the device at a current I is thus given by

$$P = IV_0(1-\eta)+I^2R \qquad (3.15)$$

where η is the junction efficiency. In many situations the junction efficiency is low and can be neglected in this equation.

Often the two components of the dissipation are of comparable magnitude, so that the power efficiency of the device is considerably lower than the junction efficiency. Furthermore, the power efficiency decreases with increasing current, and this becomes important when the pulsed operation of a device is considered. If, for example, a given mean power output can be obtained either by operating a device continuously at a current I, or by operating with a pulsed current of amplitude δI and duty cycle $1/\delta$, the former case will result in the lower power dissipation. In some applications, particularly in display systems, pulsed operation has many advantages, but these must be weighed against the increased power dissipation that will result from such pulsed operation.

3.6   Array structures

The basic principles on which electroluminescent devices are designed have been described in previous sections. These principles have been used as the basis of a wide variety of practical devices, ranging from single diodes (used, for example, as indicator lamps) to complex arrays of diodes used to make display systems. The construction of device structures that contain a single diode is a relatively straightforward problem, and needs little further comment. However, in many situations it is necessary to use a large number of diode elements closely packed to form an array. For example, a display device intended to display numbers or letters might contain 35 electro-luminescent diodes arranged in a matrix of seven rows and five columns, and there are several approaches that might conveniently provide the required format. The simplest possibility would be to take a large number of discrete devices, and package them so that the required array was obtained. This approach, although simple, does not lend itself to the construction of arrays with a large number of elements and high resolution, and does not merit a more detailed consideration. The other approaches that can be used involve mounting a number of diodes on a suitable insulating substrate or constructing the required number of diodes on one slice of semiconductor. These two approaches can lead to differences in the way in which the arrays are used, and this aspect will be considered in Chapter 6. This section is concerned with the different technologies that are used to make the arrays, and their relative merits.

3.6.1 Hybrid structures

A technique which is widely used in the semiconductor industry to make complex circuits is to mount a number of discrete devices on a ceramic substrate. This process is well suited to the construction of arrays of electro-luminescent diodes. A contact pattern is printed onto a ceramic substrate, and the diodes are bonded or soldered onto this pattern. Contacts are then made to the top surface of the diodes using a fine wire which is bonded to the diode. Using this hybrid technique, the electrical connections can be made to the diode in one of two basic configurations.

[i] *Common-anode (or -cathode) configuration.* All th eanodes (or cathodes) of the diodes can be connected to a common terminal, while separate cathode (or anode) leads are provided for each diode. This configuration is illustrated in Figure 3.20, and it can be seen that, even for a small array, the substrate connector pattern that is required can be quite complex. For an array of N diodes, the common-anode configuration requires a total of $N+1$ leads to connect the array to its drive circuits. For a large array this becomes a severe problem.

[ii] *Coordinate (or x–y) configuration.* In a coordinate-connected array, the cathodes of the diodes can be connected in rows, and the anodes can be connected in columns. This configuration is illustrated in Figure 3.21. An alternative structure, in which the cathodes of the diodes are connected in columns and the anodes are connected in rows, can also be made. The

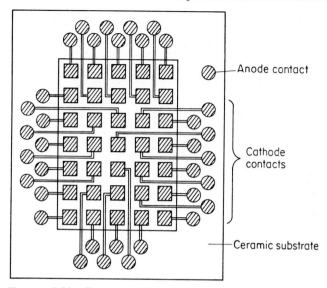

FIGURE 3.20. Contact pattern for a common-anode 7×5 array. (The contact pattern is printed on a ceramic substrate. The common-anode connection is made by a wire bonded between each diode.)

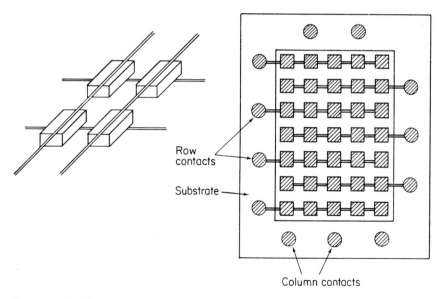

FIGURE 3.21. Contact pattern for coordinate-addressed 7×5 array. (The contact pattern for row connections is printed on the substrate. The column connections are made by wires bonded between the diodes and contact pads.)

coordinate-connected array has the advantage that the problem of connecting it to drive circuits is considerably eased, since the number of connections to the array is reduced from $N+1$, for the common-anode configuration, to approximately $\sqrt{N}$. However, the coordinate access that is required for this type of array imposes some limitations on the applications to which it can be put.

Besides mounting electroluminescent diodes on the ceramic substrate, it is also possible to mount other devices, such as the transistors required to drive the diodes, on the same substrate. This can give rise to a very convenient form of device.

3.6.2 Monolithic structure

In a monolithic structure (Figure 3.22), a large number of electro-luminescent junctions are made on one slice of semiconductor. This can be

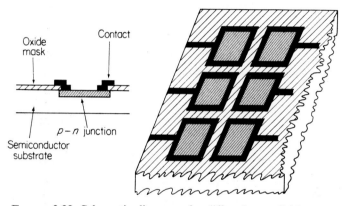

FIGURE 3.22. Schematic diagram of a diffused monolithic array

done by diffusing the junctions through a mask designed to give the required pattern of diodes. It is also possible to make a monolithic array of epitaxial diodes by confining the epitaxial growth to the required area. Although the latter approach has been shown to be experimentally feasible, it has not as yet given rise to a practical technology.

In the monolithic structure, the substrate forms a common contact to all the diodes. The other contacts, to each individual diode, are made by evaporating a contact pattern, similar to that shown in Figure 3.20, over the surface of the semiconductor. Thus the monolithic structure cannot be used to make a coordinate-connected configuration of diodes.

The monolithic approach is only feasible if it is known that the electro-luminescent devices can be produced with a high degree of uniformity; otherwise a fault in one diode will spoil the array. It is an advantage if the substrate strongly absorbs the radiation that is produced, as this minimizes the possibility of optical interaction or crosstalk between adjacent junctions. With the development of high-resolution masking and diffusion techniques, the monolithic approach can yield a very high packing density of electro-luminescent elements, and high-resolution displays can be produced at a relatively low cost. This approach is thus likely to become one of the more important ways of making alpha-numeric display devices.

3.7 The photometric properties of electroluminescent devices

The properties of a device that emits radiant energy can be measured in terms of the power that it emits, taking into account the area of the source and the angular distribution of the emitted radiation. This gives rise to the basic radiometric properties of the source, and the definitions of some of the quantities involved are indicated in Table 3.4. It is more important to specify the performance of devices that emit visible radiation in terms that relate to the effect of the radiation on the eye. To meet this requirement, a system of photometric units has evolved. For every quantity that can be defined in radiometric terms, an analogous photometric quantity can also be defined, and these definitions are discussed in the following section. The problem of relating radiometric and photometric quantities, taking into account the sensitivity of the human eye to radiation of various wavelengths, is considered in Section 3.7.2, and the measurement techniques involved are described in Section 3.7.3.

3.7.1   Basic definitions

The terms and units used in photometry have frequently given rise to confusion. This is partly because everyday terms such as 'brightness' have been used with a restricted scientific definition, and partly because of the wide range of units which have been used. An authoritive or comprehensive account of the subject will not be attempted here, and the terms and definitions that are used to specify the properties of electroluminescent devices will merely be summarized. The S.I. system of units will be used, and the definitions and units that follow are summarized in Table 3.4, together with the corresponding radiometric quantities.

[i] *Luminous intensity.* The luminous intensity of a source is a measure of the luminous power emitted by the source into a unit solid angle. The modern unit of luminous intensity is the candela (cd) and, by definition, an area of 1 cm² of a black-body radiator at the temperature of solidifying platinum has a luminous intensity perpendicular to the emitting surface of 60 candela.

[ii] *Luminous flux.* Luminous flux is the photometric equivalent of power. By definition, a source with a luminous intensity of 1 candela emits a flux

TABLE 3.4. Radiometric and photometric terms and units.

Radiometric term		Photometric term		
Quantity	Unit	Quantity	S.I. unit	Equivalent units
Radiant flux (or power)	watt (W)	Luminous flux	lumen (lm)	
Radiant intensity	watt per steradian (W/sr)	Luminous intensity	candela (cd) lumen per steradian (lm/sr)	
Radiance	watt per square centimetre steradian (W/cm² sr)	Luminance	candela per square metre (cd/m²)	1 nit 10^{-4} stilb π apostilb $10^{-4}\pi$ lambert 0·292 foot lambert
Irradiance	watt per square centimetre (W/cm²)	Illuminance	lux (lx) lumen per square metre (lm/m²)	10^{-4} phot 9.29×10^{-2} foot candle

of 1 lumen (lm) into a solid angle of 1 steradian. Thus the standard source that defines the candela also defines the lumen.

[iii] *Luminance.* The term luminance has replaced the much abused term brightness. The luminance of a source is the flux emitted per unit solid angle per unit projected source area. Luminance can thus be measured in candela per square metre or lumen per square metre steradian. Many other units have been used to measure luminance, and these are listed in Table 3.4.

[iv] *Illuminance.* The illuminance of an illuminated surface is the luminous flux density falling on the surface and is measured in lumen per square metre or lux (lx). Here again, some units that have been used to measure illuminance are listed in Table 3.4.

3.7.2   *The relationship between power and photometric units*

Quantitative measurements on visible radiation can be made in units involving power (e.g. watts) or photometric units (e.g. lumens) and, for radiation of a particular wavelength, a direct relationship between the two systems of units exists. The relationship between power and luminous flux at a particular wavelength is determined by the sensitivity of the human eye, or the luminous efficiency of radiation, at that wavelength. The relative luminous efficiency of radiation for the so-called standard observer is shown in Figure 3.23. The eye has a maximum sensitivity at a wavelength of 5560 Å, in the green region of the visible spectrum, and, at this wavelength, the luminous efficiency is 680 lm/W. Most of the devices that are available emit at longer wavelengths than this, and, as one approaches the red end of the spectrum, the sensitivity of the eye falls rapidly. In comparing the visual effect of sources of known power, this variation must be taken into account. For a source of relatively narrow spectral width, well into the visible spectrum,

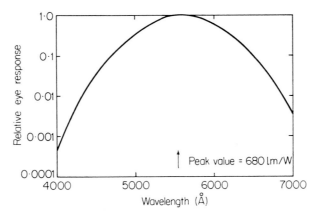

FIGURE 3.23. Relative luminous efficiency of radiation

this is a trivial problem; but, for some sources of high efficiency and broad spectral width, emission occurs in a spectral region where the sensitivity of the eye changes rapidly with wavelength. In these circumstances, an integration must be carried out to evaluate the luminous equivalent of the radiation (Figure 3.24).

Suppose that the eye-sensitivity curve is represented by $L = L_0 l(\lambda)$, where L_0 is the peak value of the luminous equivalent of radiation and has a value of 680 lm/W (at 5650 Å). The spectrum of the radiation can be represented by $P = p(\lambda)$, and is normalized so that the total power considered, $\int p(\lambda)d\lambda$, is unity. The luminous equivalent of this radiation is thus given by

$$L_0 \int l \ (\lambda)p(\lambda)d\lambda \ \text{lumen/watt.} \tag{3.16}$$

For example, the red emission from a gallium phosphide diode has a peak intensity at 6900 Å. However, the spectrum is broad and, when it is

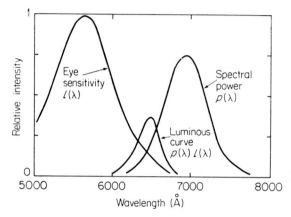

FIGURE 3.24. Evaluation of luminous equivalent of radiation (for the gallium phosphide red emission)

combined with the eye-sensitivity curve, a peak response at 6500 Å is obtained. The luminous equivalent of the radiation is approximately 15 lm/W.

3.7.3   Measurement techniques

It is not the purpose of this section to describe the way in which absolute standards of measurement have been established, but to indicate the practical techniques that are used to measure the optical properties of electroluminescent devices. In general, the accuracy of these measurements is not high, particularly for photometric properties, but, since the eye cannot detect small changes in these properties, this lack of accuracy is not important.

[i] *Power measurements.* The measurement of power in the spectral region that is important for electroluminescent devices is relatively simple. Probably the most convenient detector with which to make these measurements is a silicon *p–n* junction detector in the form of a solar cell. These have a large area, which makes it possible to measure the power emitted by the device into a defined solid angle, and the detector can be obtained calibrated over the spectral range 0·4–1 μm.

To measure the total power emitted by a device some means of integrating the power emitted in all directions must be adopted; this is commonly done using an integrating sphere. The emitting diode is placed within a sphere whose inner surface has been prepared so as to diffusely scatter any incident radiation. A small area of the sphere is removed to act as a window and allow radiation to fall onto a detector. In general, such a system does not give an absolute measurement, but must be calibrated with a source of known power.

[ii] *Photometric measurements.* In principle, the photometric properties of a device could be determined from a knowledge of its radiometric properties and the spectrum of the emission. It is, however, more convenient to make direct measurements, and to achieve this a detector whose spectral sensitivity corresponds to that of the eye is required. This can be obtained by combining a detector with a suitable filter. The accuracy of this technique depends on the accuracy with which the density of the filter and the sensitivity of the detector can be combined to match the eye sensitivity over a wide range of wavelength and sensitivity.

The most difficult of the photometric measurements is that of luminance, and most of the problems that are encountered are due to the small area of the source, which makes conventional photometric measurements impractical. An estimate of the luminance of a device can be obtained by measuring its total luminous output, the polar distribution of radiation and the source area.

An alternative method of measurement is to image the source onto a fibre-optic probe of known aperture and measure the radiation collected by the probe. If the measurements are made with a calibrated detector–filter combination, the system can be made to give a direct measurement of luminance. Instruments that are designed to make these measurements are available commercially, and give reliable results over the full range of visible wavelengths.

3.8 References

1. F. Stober, *Z. Krist.*, **61**, 299 (1925).
2. P. W. Bridgman, *Proc. Am. Acad. Arts and Science*, **60**, 305 (1925).
3. F. A. Cunnell, J. T. Edmond and W. R. Harding, *Solid State Electron.*, **1**, 97 (1960).

4. J. Czochralski, *Z. Phys. Chem.*, **92**, 219 (1917).
5. E. P. A. Metz, R. C. Miller and R. Mazelsky, *J. Appl. Phys.*, **33**, 2016 (1962).
6. J. B. Mullin, B. W. Straughan and W. S. Brickell, *J. Phys. Chem. Solids*, **26**, 782 (1965).
7. S. J. Bass, P. E. Oliver and F. E. Birbeck, *J. Cryst. Growth*, **2**, 169 (1968).
8. H. Nelson, *R.C.A. Rev.*, **24**, 603 (1963).
9. M. B. Panish, I. Hayashi and S. Sumski, *J. Quantum Electron.*, **5**, 210 (1969).
10. H. Rupprecht, in *Gallium Arsenide, Proc Intl. Symp., Reading, 1966*, IPPS, London, 1967, p. 57.
11. J. R. Knight, D. Effer and P. R. Evans, *Solid State Electron.*, **8**, 178 (1965).
12. J. J. Tietjen and J. A. Amick, *J. Electrochem. Soc.*, **113**, 724 (1966).
13. D. L. Kendall, in *Semiconductors and Semimetals*, Vol. 4 (Eds. R. K. Willardson and A. C. Beer), Academic Press, New York, 1968.
14. F. A. Cunnell and C. H. Gooch, *J. Phys. Chem. Solids*, **15**, 127 (1960).
15. P. D. Davidse and L. I. Maissel, *J. Appl. Phys.*, **37**, 574 (1966).
16. N. G. Goldsmith and N. Kern, *R.C.A. Rev.*, **28**, 153 (1967).
17. H. F. Sterling and R. C. G. Swan, *Solid State Electron.*, **8**, 653 (1965).
18. E. L. Jordan, *J. Electrochem. Soc.*, **108**, 478 (1961).
19. B. O. Seraphin and H. E. Bennett, *Semiconductors and Semimetals*, Vol. 3 (Eds. R. K. Willardson and A. C. Beer), Academic Press, New York, 1967.
20. W. N. Carr, *Infrared Physics*, **6**, 1 (1966).
21. P. Aigrain, *Physica*, **20**, 1010 (1954).
22. A. G. Fischer and C. J. Nuese, *J. Electrochem. Soc.*, **116**, 1718 (1969).
23. M. G. Holland, in *Semiconductors and Semimetals*, Vol. 2 (Eds. R. K. Willardson and A. C. Beer), Academic Press, New York, 1967.
24. H. C. Torrey and C. A. Whitmer, *Crystal Detectors*, McGraw-Hill, New York, 1948.

4

Practical Electroluminescent Devices

4.1 Introduction

The previous chapters have considered the basic theory and techniques that are the basis of electroluminescent devices. It is now appropriate to consider how these are applied to particular materials. In this discussion, emphasis will be laid on those devices that have been developed to an advanced state and are thus widely available. However, some attention will be paid to other devices which illustrate particular aspects of electroluminescent diodes and to approaches which may lead to practical devices in the future.

4.2 Gallium arsenide

The semiconductor properties of the III–V compounds were first discussed in 1952 by Welker.[1] Early work on gallium arsenide suggested that this material, with a high electron mobility and a wide band gap, might supplant germanium and silicon in conventional transistor technology. However, the exploitation of gallium arsenide to this end posed almost intractable problems, whereas silicon technology made rapid advances. This effectively ruled out the prospect that gallium arsenide would compete with silicon, and has established silicon in an impregnable technological position. However, following the continued study of gallium arsenide, it was realized that the direct band gap of this material, compared with the indirect band gap of silicon and germanium, gave it potentialities not possessed by silicon or germanium. This has resulted in semiconductor devices such as the Gunn microwave oscillator and electroluminescent diodes. Efficient electro-luminescent GaAs diodes were first reported[2] in 1962, and, since then, efficient electroluminescent gallium arsenide devices have become widely available and have found many applications.

4.2.1 The preparation and properties of gallium arsenide

The preparation of bulk single crystals of gallium arsenide now presents few problems. All the techniques that are described in Chapter 3 are applicable, although the purity of the material obtained from these techniques still leaves room for improvement. Bulk gallium arsenide has, typically, a background impurity level of about $10^{16} cm^{-3}$. This is due to contamination with silicon, and gives rise to *n*-type material with a resistivity of about 1 Ωcm.

Material with a lower carrier concentration and, correspondingly, resistivities of up to 10^8 Ωcm can be obtained by compensating the residual donors with impurities that give energy levels near the centre of the band gap. Although such high-resistivity or semi-insulating material is of considerable interest, it finds little application in electroluminescent devices.

The purest material available is produced by epitaxial techniques. In particular, material grown from a gallium solution can have a room-temperature carrier concentration of less than 10^{14}cm^{-3} and an electron mobility at room temperature that exceeds 9000 cm^2/V s, and approaches the theoretical limit set by lattice scattering.

The variation of mobility over a range of carrier concentrations is shown in Figure 4.1. In electroluminescent diodes, carrier concentrations are

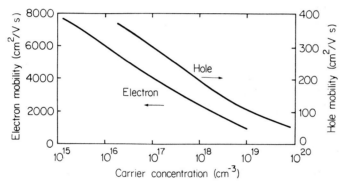

FIGURE 4.1. Electron and hole mobility in GaAs (300 K)

typically 10^{18}cm^{-3} and, in such material, the electron mobility is about 2000 cm^2/V s and the hole mobility is about 150 cm^2/V s at room temperature.

Gallium arsenide is a direct band-gap material and, of the III–V compounds, it has the largest direct band gap. This band structure, at room temperature, is shown in Figure 4.2. The variation of the band gap with temperature is shown in Figure 4.3, and can be expressed as

$$E_G = 1 \cdot 522 - 5 \cdot 8 \times 10^{-4}\, T^2/(T + 300)\ \text{electronvolt.}$$

At room temperature the band gap is 1·43 eV, so that radiation with a photon energy corresponding to band-to-band transitions will have a wavelength of approximately 9000 Å, in the near-infrared region of the spectrum.

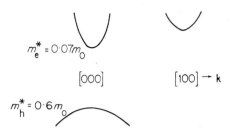

FIGURE 4.2. Band structure of GaAs
(300 K)

The experimental values of the electron and hole effective masses are

$$m_e^* = 0{\cdot}07 \ m_0 \text{ and } m_h^* = 0{\cdot}6 \ m_0.$$

The low electron effective mass means that the effective density of states in the conduction band, given by $N_c = 2(2\pi m^* kT/h^2)^{3/2}$, is only $5 \times 10^{17} \text{cm}^{-3}$ at room temperature.

In gallium arsenide, the most widely used acceptors and donors, from groups II and VI of the periodic table, respectively, introduce shallow energy levels. For example, the elements sulphur, selenium and tellurium give donor levels 4–6 meV below the conduction band, and the acceptor levels that are introduced by zinc and cadmium have ionization energies of about 20 meV.

The small impurity ionization energies and low density of states encountered in gallium arsenide cause the impurity energy levels to interact with the edges of the conduction or valence band at relatively low impurity concentrations. For example, the shallow donor levels begin to interact and form an impurity band at an impurity concentration of about 10^{17}cm^{-3},

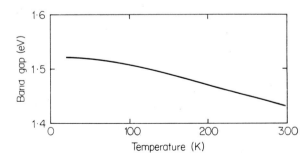

FIGURE 4.3. Variation of the band gap of GaAs with
temperature

and, at a concentration as low as $5 \times 10^{17} \text{cm}^{-3}$, the impurity levels have merged and form a 'tail' on the conduction band.

The low conduction-band density of states also has a significant effect on the optical absorption properties of gallium arsenide. Since the conduction-band density of states is low, a small concentration of electrons will fill the bottom of the band to a measurable extent. Since optical absorption can only occur when an electron is lifted into an empty state, the absorption edge in *n*-type material occurs at an appreciably higher energy than in pure material. This effect is known as the Moss–Burnstein shift. As a result of this shift, radiation which has a photon energy close to that of the band gap has a lower absorption coefficient in *n*-type material than in *p*-type material. This is of considerable importance in the design of electroluminescent diodes, as it means that the *n*-type material makes a more efficient window for junction radiation than *p*-type material.

Photoluminescence and cathodoluminescence studies on pure material at low temperatures (77 K and below) have enabled many of the recombination processes that occur in pure or lightly doped material to be studied in detail. Under these conditions, exciton effects are important, and narrow line spectra with photon energies close to the band-gap energy are observed. However, at the doping levels involved in electroluminescent *p–n* junctions, excitons do not play a significant part in recombination processes. In such material, the observed emission spectra correspond to band to impurity-level transitions, and the luminescence observed in *p*-type material is considerably more intense than that observed in *n*-type material. Such transitions are of prime importance in practical electroluminescent diodes.

The lifetime of minority electrons injected into *p*-type material can be estimated theoretically and compared with experimentally determined values. Taking a value of $10^{-9} \text{cm}^3/\text{s}$ for the recombination rate coefficient *B*, the lifetime of electrons in *p*-type material with a hole concentration of 10^{18}cm^{-3} is about 10^{-9} s, and the corresponding diffusion length of carriers is 2 μm. These values are in fair agreement with experimental values determined from measurements on gallium arsenide diodes.

4.2.2 Diffused diodes

[i] *Fabrication.* One of the simplest ways of preparing gallium arsenide electroluminescent junctions is by means of diffusion techniques. Almost invariably the acceptor zinc is diffused into *n*-type material, which may be doped with one of the group VI donors sulphur, selenium or tellurium, or the group IV element silicon which also gives a shallow donor level.

The diffusion process is usually carried out by sealing polished slices of *n*-type semiconductor in a quartz ampoule, which also contains the required diffusant. Zinc arsenide is frequently used as the diffusion source, as it gives a convenient and reproducible source of both zinc and arsenic.

Considerable effort has been devoted to optimizing the diffusion cycle to produce reliable diodes of high efficiency.

For example, in the work reported by Konnerth and coworkers,[3] 5–10 mg of ZnAs$_2$ were sealed in a 7 cm^3 ampoule. The diffusion was carried out in two stages, the first at 750°C for 4 h, followed by 45 min at 850°C. This diffusion process results in a *p–n* junction which is about 20 μm below the surface of the semiconductor.

In some circumstances it is advantageous to restrict the diffused junction to small areas of the semiconductor slice. This can easily be achieved using conventional planar techniques, as described in Chapter 3.

The diffused *p*-type layer in a gallium arsenide diode has a high carrier concentration, reaching about 10^{20}cm^{-3} at the semiconductor surface. It is easy to make ohmic contacts to this surface, but the heavily doped material strongly absorbs the radiation produced at the junction. On the other hand, as the optical absorption of the *n*-type material is relatively low, practical devices are designed so that radiation emerges from the diode through the *n*-type material.

Devices with a simple rectangular geometry or the hemispherical geometry described in Chapter 3 can be constructed. At room temperature these will have external quantum efficiencies in the range 0·2–2 per cent., the higher value referring to devices with a hemispherical or domed structure.

[ii] *Properties*. The typical emission spectrum of a gallium arsenide diffused diode is shown in Figure 4.4. The peak emission intensity occurs at a photon energy slightly lower than the band-gap energy of gallium arsenide, since the radiative processes involve impurity levels in relatively heavily doped material. The exact shape of the observed spectrum will vary somewhat, depending on the diode structure, since the absorption of radiation in the

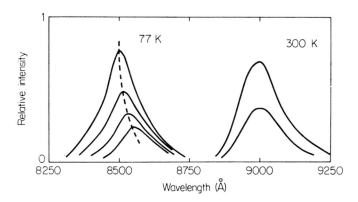

FIGURE 4.4. Spectra of gallium arsenide diffused diodes at increasing current densities

material can decrease the intensity of the higher-energy components of the emission.

At room temperature, the peak-intensity wavelength and the shape of the spectrum remain essentially constant for all operating currents. However, this is not so at lower temperatures, and it can be seen in Figure 4.4 that the emission at higher energies increases more rapidly than that at the low-energy side of the spectrum as the diode current is increased. This shift of spectrum is interpreted as the phenomenon described as 'band filling'. The electron current that is injected into the conduction band of *p*-type material effectively fills the lower energy levels in the band and, as the injection current is increased, higher energy levels must be occupied. The recombination from these higher levels results in a progressive shift of the diode spectrum to higher energies.

As is to be expected with a diode made from a direct band-gap material with short minority-carrier lifetimes, the response of a gallium arsenide diffused diode, when subjected to a pulse of current, is rapid. The peak emission intensity is reached with a rise time of about 10 ns, and the emission decays at a similar rate when the drive current is removed.

4.2.3   Epitaxial silicon-doped diodes

An important advance in the development of gallium arsenide electro-luminescent devices was reported in 1966 by Rupprecht.[4] This involved the use of silicon as both the donor and acceptor impurity in gallium arsenide diodes made using epitaxial techniques. Such diodes, in which one element is used to provide both donor and acceptor levels, are often called amphoterically doped diodes.

[i] *Preparation.* When gallium arsenide doped with silicon is grown from a stoichiometric melt of gallium arsenide, the silicon gives rise to shallow donor levels. These levels are often the residual levels that are observed in pure material. On the other hand, when gallium arsenide is grown from a gallium solution doped with silicon, the silicon atoms can be incorporated in the GaAs lattice on either Ga or As sites. On the gallium sites, silicon is a shallow donor, but, on the arsenic sites, silicon produces acceptor levels with an ionization energy of about 30 meV. This simple picture can explain the main features of the donor–acceptor behaviour of silicon. However, there is some evidence that the situation is more complicated than this, with atomic complexes playing an important part.

Amphoterically doped diodes are grown epitaxially, from a silicon-doped solution of GaAs in gallium, onto an *n*-type substrate. While the growth temperature is above 750°C, *n*-type material is grown. Below this temperature, the grown layer is *p* type. Thus, as growth proceeds, a diode structure is obtained. At the transition temperature of 750°C, closely compensated *p*-type material is grown. This material is designated p^0, and the structure can be described as a p–p^0–n structure, in contrast to the p–n structure of a diffused diode.

The radiation-emitting region in a p–p^0–n structure is very wide. Typically, radiation is produced in a volume which extends 50 μm from the junction, compared with a width of 2 μm in other structures.

[ii] *Properties.* At room temperature, the peak emission intensity occurs at a wavelength of 9300–9700 Å, depending on the silicon concentration,[5] compared with 9000 Å for diffused diodes (Figure 4.5). This longer wave-length and the broad emitting region result in diodes with a considerably reduced internal absorption loss, so that the efficiency of amphoterically doped diodes is considerably greater than that of diffused diodes. For simple slab structures, external quantum efficiencies of 6 per cent. have been reported, while a hemispherical structure can have a quantum efficiency as high as 30 per cent. at room temperature. Efficiencies a factor of two higher than these are observed at liquid-nitrogen temperatures. These high external

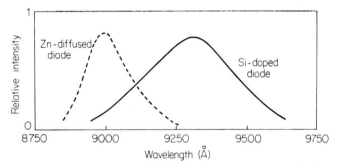

FIGURE 4.5. Spectra of gallium arsenide Si-doped diodes
(300 K)

efficiencies imply that the internal quantum efficiency of these devices is probably as high as 50 per cent. at room temperature, and approaches 100 per cent at liquid-nitrogen temperatures.

Another important difference between amphoteric and diffused diodes is their response time. The response of a p–p^0–n diode to a pulse of current is shown in Figure 4.6; the diode has a rise time of about 200 ns. This relatively slow response imposes stringent limitations on the applications of amphoteric diodes, as they cannot be modulated efficiently at frequencies much greater than 1 MHz.

4.2.4 Phosphor-coated diodes

Some phosphors, which are known as up-converting or anti-Stokes phosphors, can absorb two photons at one wavelength and emit a single

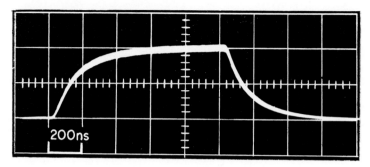

FIGURE 4.6. Transient response of gallium arsenide Si-doped diodes
(300 K)

photon at a shorter wavelength. In 1968, Galginaitis and Fenner[6] showed that this could give rise to an approach to the problem of making 'visible' electroluminescent devices. In this work, they used a phosphor to 'up convert' the infrared radiation from a gallium arsenide diode into visible radiation.

The earliest devices used a phosphor, consisting of an LaF_3 host lattice activated with erbium and sensitized by ytterbium, which was coated onto the surface of a silicon-doped gallium arsenide diode.

The phosphor absorbs radiation at 9700 Å, and, although the emission from the gallium arsenide diode does not exactly match this absorption, the infrared radiation extends to a long enough wavelength to stimulate the phosphor with a useful efficiency. An outline of the energy-level scheme involved is shown in Figure 4.7.

The first infrared photon is absorbed in raising the Yb^{3+} ion to the $^2F_{5/2}$ state, and the energy is then transferred to the Yb^{3+} ion, raising it to the $^4I_{11/2}$ state. Subsequent absorption and transfer processes give the $^4F_{7/2}$ and $^4G_{7/2}$ states of the Er ion. These states decay with the emission of radiation at wavelengths between 3800 and 6600 Å.

Subsequent work has shown that higher conversion efficiencies can be obtained using host lattices such as $BaYF_5$, YOCl, Y_3OCl_7 or YF_3, and that the choice of the host lattice determines, to some extent, which emission processes are favoured. By using the rare-earth elements thulium (Tm) or holmium (Ho) in place of erbium, a range of emission wavelengths can be obtained. Some of the systems that have been studied are listed in Table 4.1, and a review of the properties of these systems is given in References 7 and 8.

In order to improve the efficiencies of devices, it is necessary to improve the match between the diode emission and the phosphor absorption.

This may be done by using heavily doped GaAs diodes, which emit at wavelengths close to 9700 Å and can have an efficiency as high as 10 per cent. (see Section 4.2.3). With this approach, the broad spectrum of the emission

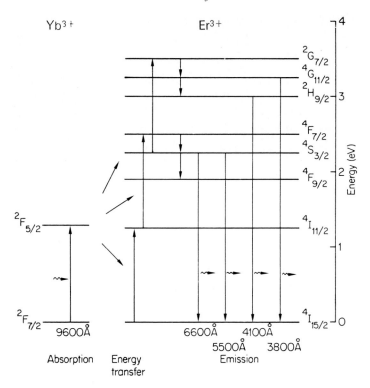

FIGURE 4.7. Simplified diagram of absorption and emission paths in $Er^{3+}-Yb^{3+}$ phosphors

TABLE 4.1. Anti-Stokes phosphors for infrared-to-visible-conversion electroluminescent devices.

Host lattice	Rare-earth activator	Dominant emission (Å)
YOCl : Yb	Er	6600
Y_3OCl_7 : Yb	Er	5500
YF_3 : Yb	Er	5500
$BaYF_5$: Yb	Er	5500
YF_3 : Yb	Tm	4700

limits the amount of radiation that can be absorbed by the phosphor absorption band.

An alternative approach is to use indium phosphide diodes to stimulate the phosphor. Although these emit at the appropriate wavelength, they have not been highly developed and have efficiencies of only about 1 per cent.

Taking into account the spectral width of the two diode emissions, the pumping efficiency of a 1 per cent. InP diode is comparable to that of a 5 per cent. Si-doped GaAs diode, so that both approaches are feasible and may lead to useful device improvements.

For a two-photon excitation process, the intensity of emission increases as the square of the excitation intensity. Thus, to obtain high efficiencies, the semiconductor diodes must be operated at high power densities, and it is important that the phosphor should be as close to the infrared source as possible. For this reason, the diode structure must be carefully designed, and the phosphor is coated directly onto the diode using a suitable binding medium.

Some emission processes depend on three-photon stimulation, and the emission intensity then increases as the cube of the input current. If both two- and three-photon processes can occur, it is possible to change the colour of the emission by varying the drive current.

Using gallium arsenide diodes with 10 per cent. efficiency at 9300 Å, and operating at a current density of about 100 A/cm^2, overall efficiencies of 1 per cent. (red), 0·1 per cent. (green) and 0·01 per cent. (blue) have been obtained.[9] These efficiencies are approaching those of other red- and green-emitting devices, and are better than other blue-emitting devices. However, the current densities are an order of magnitude greater than those which are used in most diodes, and this is a severe limitation.

4.3 Gallium arsenide–phosphide and other III–V ternary compounds

The successful exploitation of gallium arsenide to produce efficient infrared-emitting devices naturally leads to a consideration of ways in which high efficiencies might be obtained from devices emitting visible radiation. The obvious extension of the gallium arsenide work is thus to seek a material with a direct band gap that is sufficiently wide to give visible recombination radiation. Taking the long-wavelength limit of the visible spectrum to be 7000 Å implies that a band gap of 1·8 eV or greater is required. However, as has already been noted, gallium arsenide is the widest direct band-gap material of the III–V compounds, and to extend this gap it is necessary to utilize the ternary III–V compounds of the type $III^A_x III^B_{1-x} V$ or $III V^A_x V^B_{1-x}$. The band structure of these materials was reviewed in Chapter 2, and it was concluded that any of the systems InP–GaP, GaAs–GaP, GaAs–AlAs, InP–AlP or InAs–AlAs would have a suitable band gap.

D

To narrow the choice of materials still further, the problems of making material of the required composition, and the degree of purity and crystal perfection required in semiconductor devices, must be considered.

In materials of the type III $V_x^A V_{1-x}^B$, it is to be expected that the best crystal structure would be obtained if the V^A and V^B atoms had similar tetrahedral radii. Similar considerations apply to $III_x^A III_{1-x}^B V$ materials, in which the proportion of the III^A and III^B elements are varied. Alternatively, regarding the ternary materials as mixtures of two binary compounds with a common element, a system in which the binary compounds have a similar lattice parameter should give the best material.

The tetrahedral radii of the III–V elements and the lattice parameters of their compounds are given in Table 4.2. From the data given here, it can

TABLE 4.2. Tetrahedral radii of group III and group V elements and lattice parameters of the III–V compounds (values in Å).

Group III element and tetrahedral radius	Group V element and tetrahedral radius		
	P 1·10	As 1·18	Sb 1·36
Al 1·26	AlP 5·42	AlAs 5·62	AlSb 6·14
Ga 1·26	GaP 5·45	GaAs 5·65	GaSb 6·10
In 1·42	InP 5·87	InAs 6·06	InSb 6·48

be seen that the GaAs–AlAs system is almost perfectly matched, followed by the GaAs–GaP and InP–GaP systems. In Table 4.2, which is arranged with the rows (Al Ga In) and columns (P As Sb) in order of increasing atomic number, the constituents of these three systems are in adjacent rows or columns.

Since the binary compounds GaAs and GaP have both been extensively studied, it is not surprising that the $GaAs_x P_{1-x}$ system has been the most studied ternary system.

A wide range of gallium arsenide–phosphide devices has been developed, and most of this section will be devoted to this material. However, since the ternary materials $Ga_x Al_{1-x} As$ and $Ga_x In_{1-x} P$ show considerable promise and have some potential advantages over $GaAs_x P_{1-x}$, these will also be briefly discussed.

4.3.1 The preparation and properties of gallium arsenide–phosphide

Bulk crystals of the ternary alloys cannot be made by the techniques that are used to grow single crystals of GaAs, GaP, etc. Gallium arsenide–phosphide is usually prepared by an epitaxial technique, using gallium arsenide as the substrate. The hydride system, described in Section 3.3.2, is particularly convenient, since, by using the hydrides AsH_3 and PH_3 and varying their relative proportions, the composition of $GaAs_xP_{1-x}$ grown can easily be controlled. To grow material of good crystal structure on a substrate that has a different composition and lattice parameter, it is necessary to control the growth so that the first layers grown match the substrate. The composition of the grown material can then be graded to that required for the device.

The band structure of the $GaAs_xP_{1-x}$ alloys was briefly reviewed in Chapter 2, and is shown in more detail in Figure 4.8. Although it is often assumed that the band structure of the ternary alloys can be deduced by a linear interpolation, with composition, between the structures of the binary constituents, this is not always a very accurate assumption; but it is adequate for this material. In the GaAs–GaP system, the transition from direct to indirect band-gap material occurs at a composition of $GaAs_{0.55}P_{0.45}$ with

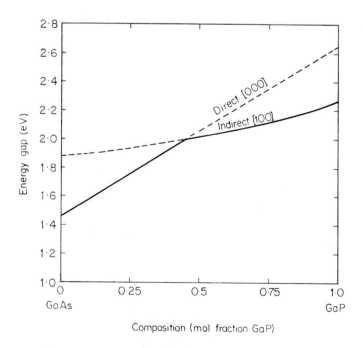

FIGURE 4.8. The band structure of $GaAs_xP_{1-x}$ materials, showing the direct–indirect transition point (300 K)

a band gap of 2·0 eV at room temperature. These parameters vary slightly with temperature and, at 77 K, the transition occurs at a composition with a slightly lower arsenic content.

Impurity levels also affect the apparent direct–indirect transition. Because the effective mass in the [100] conduction band (m_{100}^*) is considerably greater than that in the [000] conduction band (m_{000}^*), donor levels associated with the [100] band are deeper than those associated with the [000] band. Thus, compared with pure material, the effective direct–indirect transition in *n*-type material can be expected to occur at an alloy composition that is displaced somewhat towards the GaAs end of the system.

The electrical properties of the direct band-gap materials are similar to those of gallium arsenide. For example, the room-temperature electron mobility lies in the range 5000 to 2000 cm^2/V s for electron concentrations of 10^{16}–10^{18}cm^{-3}. On the indirect side of the system, mobilities are about an order of magnitude lower, corresponding to those in gallium phosphide.

4.3.2 *Gallium arsenide–phosphide diodes*

[i] *The formation of p–n junctions in* GaAs$_x$P$_{1-x}$. *p–n* junctions can be formed in gallium arsenide–phosphide either by changing the dopant during the vapour-phase growth of the material, or by using diffusion techniques.

The use of epitaxial techniques enables complex device structures to be grown.[10] For example, the junction can be formed at the optimum doping level for high efficiency. Material adjacent to the junction can have a different composition to minimize its optical absorption, and the contact region can be highly doped to give low-resistance contacts. A typical junction structure grown in this way is shown in Figure 4.9.

Diffused junctions are formed using techniques similar to those used with gallium arsenide,[11] and again zinc is the diffusant which is almost

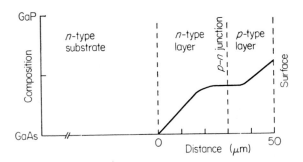

FIGURE 4.9. Composition profile of a GaAs$_x$P$_{1-x}$ diode, grown by vapour-phase epitaxy on a GaAs substrate

invariably used. Planar technology has been extensively developed, and has proved to be extremely useful in making monolithic arrays of diodes for display purposes and in improving the reliability of devices.

[ii] *The properties of GaAs$_x$P$_{1-x}$ junctions*. The electroluminescent efficiency of gallium arsenide–phosphide devices changes markedly with the transition from direct to indirect band-gap materials. With direct band-gap materials, efficiencies of 2 per cent. or more are obtained, comparable to those obtained in gallium arsenide, while the indirect band-gap materials give efficiencies

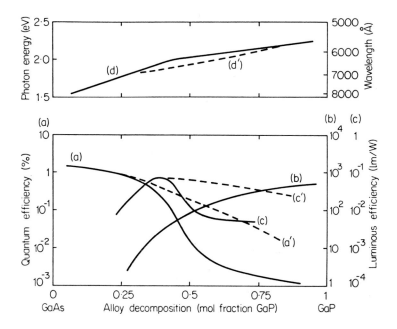

FIGURE 4.10. The properties of GaAs$_x$P$_{1-x}$ diodes: (a) quantum efficiency, (b) luminous efficiency of radiation (eye sensitivity), (c) luminous efficiency of diode and (d) photon energy and wavelength of emission; curves (a'), (c') and (d') represent results obtained with nitrogen doping

of less than 0·01 per cent. The typical variation of efficiency with composition is shown in Figure 4.10. Results such as these can be explained by considering the detailed band structures of the materials and the relative electron populations of the conduction-band minima when electrons are injected into the material.[12] It is important to note that the density of states in the [100] band is much greater than that in the [000] band. Thus, when the two bands are level, corresponding to the direct–indirect transition, the majority of electrons injected into the conduction band of *p*-type material will occupy the [100] minima and recombine non-radiatively. To obtain efficient radiative recombination, it is thus necessary to use material that lies appreciably on the arsenic-rich direct band-gap side of the composition $GaAs_{0.55}P_{0.45}$.

Since gallium arsenide–phosphide devices are required for use in the visible region of the spectrum, it is important to design a device for optimum luminous efficiency. In the region of the spectrum that is of interest here, the sensitivity of the eye decreases by a factor of ten over a wavelength change of 400 Å, corresponding to a change in photon energy of 0·1 eV. The optimization of material composition and device efficiency is thus important.[13] Fortunately, the variations of eye sensitivity and device efficiency tend to compensate each other, and their optimization is not as critical as might be expected. The variation of luminous efficiency with composition for a series of diodes is shown in Figure 4.10. From these results, the optimum material composition is $GaAs_{0.6}P_{0.4}$, giving a device efficiency of 0·2 per cent. at a wavelength of 6500 Å. At this wavelength, the luminous efficiency of radiation is about 50 lm/W.

Recent experimental work[14] has shown that it is possible to increase the efficiency of $GaAs_xP_{1-x}$ diodes by doping the *n*-type material with nitrogen in addition to the usual donor. This technique was first used in gallium phosphide diodes, and is discussed in Section 4.4.3.

For a given composition of $GaAs_xP_{1-x}$, the effect of nitrogen doping is to increase the diode efficiency, but at the same time decrease the photon energy of emission, as shown in Figure 4.10.

The net result is that the gain in efficiency is more significant than the lowered luminous efficiency of the radiation. Thus diodes with a luminous efficiency comparable to red-emitting devices can be made at wavelengths as short as 5600 Å. This brings the emission into the yellow region of the spectrum.

4.3.3   *Gallium arsenide–phosphide devices*

At present, the majority of red-emitting electroluminescent devices that are available are based on zinc-diffused gallium arsenide–phosphide diodes. As will be seen, these diodes have not shown such high efficiencies as gallium phosphide diodes, but the advanced state of the technological development of gallium arsenide–phosphide has, to a large extent, outweighed this disadvantage.

The planar technology that has been developed with gallium arsenide–phosphide is particularly suited to the fabrication of monolithic arrays of diodes, particularly where high resolution is required. The hybrid technology, in which a number of discrete diodes are bonded to a ceramic substrate, is also widely used to form gallium arsenide–phosphide devices. Some examples of devices made by these technologies are illustrated in Section 4.8.

4.3.4 *Gallium–aluminium arsenide and gallium–indium phosphide*

[i] *Gallium–aluminium arsenide.* The binary constituents of the $Ga_xAl_{1-x}As$ system, i.e. GaAs and AlAs, have lattice spacings that are very closely matched, and this has favoured the growth of good-quality material over a range of compositions.

Gallium–aluminium arsenide is usually grown by a liquid-phase epitaxial technique using a gallium arsenide substrate. The phase diagram of this system indicates that the distribution coefficient of aluminium between Ga–Al–As solid and liquid phases is much greater than unity, so that a small concentration of aluminium in the melt will produce a much larger concentration of aluminium in the solid. For example, with a growth temperature of 1000°C, a solution containing approximately 0·8 per cent. of aluminium atoms will give a solid of composition $Ga_{0.6}Al_{0.4}As$.

The composition $Ga_{0.6}Al_{0.4}As$ corresponds to the transition between direct and indirect band-gap material and has an energy gap of 1·9 eV. This is a somewhat lower energy than in the GaAs–GaP system, so that gallium–aluminium arsenide devices will not extend the spectral range of visible electroluminescence to shorter wavelengths than those obtained from gallium arsenide–phosphide. However, the relative ease with which the material can be grown is a considerable advantage, and devices with an efficiency of 0·15 per cent. at 1·89 eV have been obtained. The wavelength of the emission is approximately 6550 Å, with a luminous efficiency of 40 lm/W, giving results that are comparable to those of gallium arsenide–phosphide. At longer wavelengths efficiencies of 4 per cent. (6950 Å) to 0·3 per cent. (6700 Å) have been reported.[15]

The silicon amphoteric doping technique, which has been used with considerable success to give high-efficiency gallium arsenide diodes, can also be used with gallium–aluminium arsenide.[16] *p–n* junctions can be grown by a single-stage epitaxial process in which silicon-doped gallium–aluminium arsenide is grown from a melt. As with gallium arsenide, the initial growth gives *n*-type material, and *p*-type material is formed as the growth temperature drops.

Using these epitaxial techniques, diodes with an efficiency of 2 per cent. have been made in material with a band gap of 1·4–1·6 eV, but, as the band gap increased further, the efficiency dropped to 0·02 per cent. at 2 eV. As with similar GaAs devices, the wavelength of emission is about 400 Å longer than that corresponding to the band-gap energy of the material. These results show efficiencies that are only comparable to the best diffused diodes, and should thus be capable of considerable improvement. However, since the emission wavelength is longer than that of diffused diodes, this approach will not lead to useful improvements in the visible spectral range, but will give useful devices over a range of near-infrared wavelengths.

[ii] *Gallium–indium phosphide.* The various reported measurements of the

band structure of the GaP–InP system show some discrepancies in the composition at which the direct–indirect transition occurs. However, it appears that the maximum direct band gap is approximately 2·2 eV, and this offers considerable advantages over the other ternary materials that have so far been explored. The technology of this material has not advanced very far because difficulties have been encountered in growing suitable material and, at present, only a few studies on electroluminescent devices have been published. These have reported efficiencies of about 0·1 per cent. at a wavelength of 6000 Å.[17]

4.4   Gallium phosphide

Although gallium phosphide is an indirect band-gap material, and might thus not be expected to show efficient luminescence, it has been known for some time that useful devices can be made in gallium phosphide. In fact, gallium phosphide devices were among the first electroluminescent diodes to find viable application.

The band gap of gallium phosphide is 2·2 eV, and radiative recombination can give emission in the green or longer wavelength regions of the spectrum. At present, two transitions are widely exploited, one giving rise to green and one to red electroluminescence.

The red electroluminescence was the first to be exploited. Early work by Starkiewicz and Allen[18] showed that, when zinc and oxygen are used to dope GaP, a luminescent centre is formed which gives rise to a radiative transition with a photon energy of 1·8 eV (6900 Å).

The green transition corresponds closely to the band-gap energy. There are several transitions that could, in principle, give such a transition, but recent work has shown that the most efficient devices are obtained using material that is doped with nitrogen.

4.4.1   The preparation and properties of gallium phosphide

Gallium phosphide can be synthetized by the reaction of phosphorus with gallium, but this technique gives material that is of poor crystalline quality and unsuitable for device development. Early device work relied on small crystals in the form of platelets, which could be obtained by the freely nucleated crystallization of GaP from a gallium solution. This technique can yield platelets with linear dimensions of up to 1 cm, and the purity and electrical properties of this material are good. However, material in this form does not make an economic basis for device development.

The Czochralski technique has, in recent years, been successfully applied to the growth of gallium phosphide.[19] Growth is carried out under a flux of boric oxide, and single-crystal ingots 2 cm in diameter and 8 cm long can be grown. At present, this material usually does not show the electrical

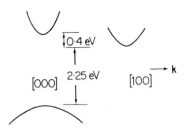

FIGURE 4.11. Band structure of
GaP (300 K)

quality required for electroluminescent devices. The exact deficiencies have
not yet been identified, but it may well be that lattice defects play a part in
reducing the efficiency of radiative recombination in the material.

Epitaxial techniques have proved very successful in growing gallium
phosphide. In early work, GaP was grown by a vapour-phase technique on
gallium arsenide substrates. Now that single-crystal material grown by the
Czochralski technique is available, material can be grown by vapour- and
liquid-phase techniques on gallium phosphide substrates.

The band structure of gallium phosphide is shown in Figure 4.11. The
[100] indirect band gap has a value of 2·25 eV at room temperature, and
the [000] conduction-band minimum lies 0·4 eV higher. The variation of
band gap with temperature is shown in Figure 4.12, and can be represented
as

$$E_G = 2{\cdot}338 - 6{\cdot}2 \times 10^{-4} T^2 / (T + 460) \text{ electronvolt.}$$

The electron and hole mobilities of gallium phosphide are shown in
Figure 4.13 for a range of carrier concentrations. At a carrier concentration
of 10^{18}cm^{-3}, which is typical of the material used in devices, electron

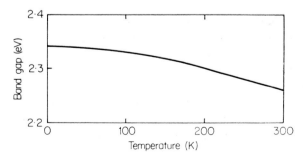

FIGURE 4.12. Variation of the band gap of GaP with
temperature

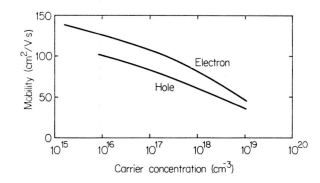

FIGURE 4.13. Electron and hole mobility in GaP (300 K)

mobilities are about $100 \text{ cm}^2/\text{V s}$ and hole mobilities are about $500 \text{ cm}^2/\text{V s}$.

In gallium phosphide, the group VI elements S, Se and Te give shallow donor levels with ionization energies of about 0·1 eV, and silicon also gives a shallow donor level of similar energy. Oxygen gives a donor level 0·8 eV below the conduction band. The group II elements behave as acceptors with ionization energies in the range 0·05–0·1 eV. As in gallium arsenide and gallium arsenide–phosphide, zinc is almost universally used to dope *p*-type material, although cadmium is sometimes used to investigate the effect of a different, but known, acceptor ionization energy.

The recombination processes that occur in GaP have been extensively studied, and probably more attention has been paid to this material than any other. Several extensive reviews of this work have been published.[20] For the present purposes it is necessary to consider in detail only the two processes that give rise to practical devices.

[i] *Red luminescence.* The red luminescence in gallium phosphide is due to an energy level that arises when both zinc and oxygen are present in the form of a Zn–O complex. As has already been noted, when zinc is introduced into the GaP lattice, it substitutes for gallium and gives an acceptor level 0·06 eV from the valence band. In a similar way, oxygen substitutes for phosphorus and produces a deep donor level that is 0·8 eV below the conduction band. These energy levels are observed when the zinc and oxygen atoms occupy lattice sites that are too far apart for any mutual interaction to take place. However, when the zinc and oxygen atoms occupy adjacent lattice sites, the resulting Zn–O complex has a net electron attraction and thus behaves as an electron trap that has an energy level 0·3 eV below the conduction band.

If an electron is injected into the conduction band of *p*-type material, it can be trapped on the Zn–O level. The Zn–O complex is now negatively charged, and can trap a hole to form an exciton which is bound to the Zn–O centre. The exciton can now decay radiatively with the emission of a photon whose energy is given by $E_G - E_{Zn-O} - E_x$, where the energy levels are defined in Figure 4.14.

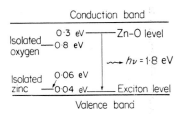

FIGURE 4.14. Energy levels in Zn–O doped GaP. The transition giving rise to efficient red emission is indicated

Alternatively, the trapped electron can decay by recombining with a hole at an acceptor level, but this process is believed to be of less significance at room temperature.

[ii] *Green luminescence.* Among the many luminescent processes that can give rise to radiation with a photon energy close to the band-gap energy of gallium phosphide, that involving the isoelectronic impurity nitrogen is the most important in the context of practical electroluminescent devices emitting in the green region of the spectrum.

In an indirect band-gap semiconductor such as gallium phosphide, the probability of a band-to-band transition is small. However, when nitrogen is introduced into the GaP lattice, it substitutes on phosphorus sites and gives rise to a relatively efficient radiative recombination process. The outer electronic structure of nitrogen is similar to that of phosphorus, but there is considerable difference between the electronic core structures of these atoms, and this perturbation gives rise to an electron trap close to the conduction band. An energy level produced in this way is known as an isoelectronic level and can, in effect, give energy states that are not confined to the [100] point.

An electron that is injected into *p*-type material containing nitrogen can be trapped on the isoelectronic level. The charged centre can now interact with a hole to give rise to a radiative decay process. This process can be relatively efficient at room temperature, and results in a photon with an energy of 2·2 eV.

[iii] *Yellow luminescence.* If the concentration of isoelectronic nitrogen atoms is increased to $5 \times 10^{19} cm^{-3}$ or more, recombination processes that

involve pairs of nitrogen atoms on adjacent lattice sites become important.[21] Here again the recombination is an efficient radiative process and a photon with an energy of 2·1 eV is emitted.

4.4.2   Red-emitting GaP diodes

[i] *The formation of p–n junctions.* The earliest practical GaP electro-luminescent diodes were obtained by alloying tin into *p*-type Zn–O doped gallium phosphide platelets.[18] This process gave diodes with an efficiency of 0·01–0·1 per cent., but the process was not very well suited to the economic production of devices, and epitaxially grown junctions have almost entirely supplanted these early devices.

The processes involved in the economic production of efficient devices have evolved through a number of stages. Diodes with an efficiency of 1 per cent. were made by epitaxially growing an *n*-type layer on solution-grown Zn–O-doped platelets. Although this process eliminated the use of alloyed junctions, the use of platelets as starting material is very inconvenient.

The obvious approach to the problem is to use bulk single-crystal material, such as is grown by the Czochralski technique, and grow a junction on this. However, at present this has not proved to be a very reproducible process, presumably owing to some unidentified impurity or lattice defects in the bulk material.

Probably the most successful techniques are based on the work described by Saul, Armstrong and Hackett.[22] In this approach, a tellurium-doped layer is grown on an *n*-type substrate, and the junction is formed by growing a Zn–O-doped *p*-type layer. Both epitaxial layers are grown from a solution of gallium phosphide and gallium at a temperature of about 1000°C. The solution for the *n*-type layer is doped with tellurium, and that for the *p*-type layer is doped with ZnO and Ga_2O_3. The structure can be grown on bulk single-crystal gallium phosphide substrates, and a typical doping profile is shown in Figure 4.15. This process lends itself to the economic production of diodes, and efficiencies of about 1 per cent. can be obtained.

The efficiency of such diodes can be increased considerably by annealing the junction at a temperature of 500°C for 10–20 h. The annealing process can conveniently be carried out as a final stage of the last epitaxial growth, while the slice is still immersed in gallium. It is believed that this process favours the association of zinc and oxygen atoms to form the Zn–O complex required in the recombination process, and also reduces concentration of non-radiative centres in the structure. Devices made in this way can have efficiencies up to 7 per cent., although a somewhat lower figure (about 2 per cent.) is more representative of typical diodes.

Considerable efforts have been made to produce diodes by diffusion techniques. Junctions made by diffusing zinc into tellurium–oxygen-doped material have been more successful than other approaches, and efficiencies

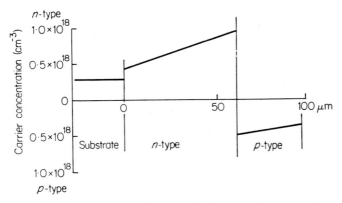

FIGURE 4.15. Doping profile through the epitaxial layers of a
Zn–O doped GaP diode

of 1 per cent. have been obtained. Although these efficiencies are considerably
below those obtained by epitaxial techniques, they are adequate for many
purposes and open the way to the exploitation of a planar monolithic
technology in gallium phosphide.

[ii] *Properties of red-emitting diodes.* Gallium phosphide diodes show the
typical current–voltage characteristics that are obtained when space-charge
recombination and diffusion currents are both significant. Thus the emission
from a diode increases approximately as the square of the current density at
low current densities ($0\cdot1$ A/cm^2) and then becomes linear. If the current
density is increased to about 10 A/cm^2, the output becomes sublinear,
tending to a square-root dependence on current (Figure 4.16). Thus decreases

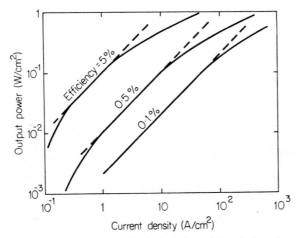

FIGURE 4.16. Output characteristics of Zn–O doped
GaP diodes (300 K)

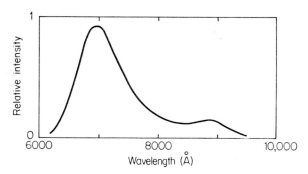

FIGURE 4.17. Spectrum of a Zn–O doped GaP diode
(300 K)

in efficiency can be associated with a quasi-saturation of the radiative recombination path through the Zn–O centres, and detailed studies of these characteristics enable an estimate of the concentration of Zn–O centres near the junction to be made.

The spectrum of red emission from a gallium phosphide diode is shown in Figure 4.17. The peak of the emission occurs at 6900 Å, and the eye is not very sensitive at this wavelength. The visual effect of the radiation is mainly due to the shorter-wavelength component of the emission. By combining the spectrum with the eye-sensitivity curve, it can be shown that the peak visual stimulation occurs near 6500 Å, and that the luminous efficiency of the radiation is 15 lm/W.

The transient response of a diode output when subjected to a pulse of current is shown in Figure 4.18. This response shows rise and fall times of 100 ns. These times are fast when compared with the speed of response of the human eye, but may be significant in situations in which the diode is

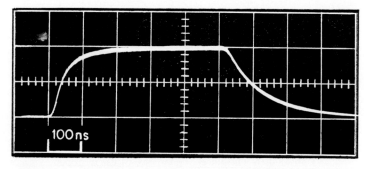

FIGURE 4.18. Transient response of a Zn–O doped GaP diode (300 K)

operated at high modulation rates, such as may occur in some scanned display systems.

The self absorption of the red emission is small, so that in many devices radiation is emitted over a solid angle of 4π steradians. This must be considered when estimating the brightness of a device if its efficiency and power output are based on measurements carried out in an integrating sphere. Total-power and efficiency measurements on red-emitting gallium phosphide devices will give these devices an apparent advantage when compared with devices in which self absorption is high, and which thus emit only over a solid angle of 2π steradians. In many situations, it is not possible to make use of the full power emitted into a 4π steradian solid angle by the gallium phosphide device, and as much as half the power may be wasted.

4.4.3 Green-emitting gallium phosphide diodes

[i] *The formation of p–n junctions.* The formation of green-emitting diodes requires the careful exclusion of oxygen to eliminate the centres that give rise to red electroluminescence and the introduction of nitrogen to provide the required isoelectronic centres. Using the techniques described by Logan, White and Wiegman,[23] diodes with efficiencies of up to 0·6 per cent have been made. These diodes were grown by liquid-phase epitaxial techniques in which p-type material, doped to a zinc concentration of 10^{18}cm^{-3}, was grown on a sulphur-doped n-type substrate with a carrier concentration of 5×10^{16}cm^{-3}. Both growth processes were carried out under a flow of hydrogen. Oxygen was eliminated from the system by passing the gas over a sodium trap, and nitrogen doping was achieved by introducing ammonia into the gas flow at a concentration of 0·06 per cent.

Green-emitting diodes have also been made by diffusion techniques. By diffusing zinc into n-type material containing nitrogen centres, efficiencies of 0·05 per cent. have been obtained. Although this is an order of magnitude lower than the best achieved by epitaxial techniques, the efficiency is reproducible, and adequate for many purposes. Furthermore, it enables devices based on a planar technology to be developed.[24]

[ii] *Properties of green-emitting diodes.* At low forward biases, the current–voltage characteristic of nitrogen-doped diodes indicated that the current is due to space-charge recombination, and the radiative output increases approximately as the square of current. At higher values of bias, a linear output-versus-current region can be seen. Unlike red-emitting, diodes, this linear region of constant diode efficiency continues throughout the range of practical drive conditions, and the recombination path does not saturate.

The spectrum of the green emission is shown in Figure 4.19. This shows a peak emission at 2·22 eV at room temperature. The corresponding wavelength is 5650 Å, which is very near the peak of the eye-sensitivity curve,

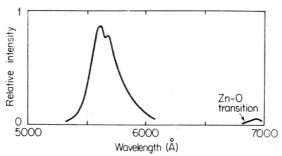

FIGURE 4.19. Spectrum of a nitrogen-doped GaP diode
(300 K)

and the luminous efficiency of the radiation is 600 lm/W. In practice, diodes may show a red component of the spectrum owing to oxygen contamination, and the appearance of such diodes can vary from a pure green to a yellow–green.

4.4.4   Yellow-emitting gallium phosphide diodes

It has been reported[21] that diodes made by diffusing zinc into heavily nitrogen-doped gallium phosphide have shown an efficiency of 0·1 per cent. with a photon energy of 2·1 eV. This photon energy corresponds to a wavelength of 5900 Å, which is in the yellow region of the visible spectrum.

4.4.5   Gallium phosphide devices

Gallium phosphide diodes can be encapsulated and packaged in a number of ways, and the high luminous efficiency that can be obtained for red- and green-emitting devices makes them very suitable for low-power indicator lamps.

Gallium phosphide diodes can be built into arrays on ceramic substrates, and similar technologies can give either red- or green-emitting devices.

At present, the technology of gallium phosphide is somewhat less advanced than that of gallium arsenide–phosphide, and planar arrays of GaP diodes have not been so widely developed. However, there seems little doubt that such a technology can be developed and is likely to become important, particularly for green-emitting devices, where this seems to be the most viable approach.

Some examples of gallium phosphide devices are illustrated in Section 4.8.

4.5   Silicon carbide[25]

Silicon carbide was one of the first semiconductors from which electro-luminescence was observed, but the chemical nature of the material makes it very difficult to prepare, and the development of electroluminescent devices has not advanced very far. However, since the band gap of the material is between 2 and 3 eV, depending on the crystal form, it might produce devices throughout the range of the visible spectrum, and for this reason deserves brief consideration.

Silicon carbide has an indirect band gap and might thus be expected to show properties analogous to those of gallium phosphide. Although such analogies can be drawn, and similar radiative processes identified, the technology of SiC has not developed to the same extent as that of GaP and the efficiency of the devices obtained are about two orders of magnitude lower.

4.5.1 The preparation and properties of silicon carbide

Silicon carbide does not possess a liquid phase, and must be prepared by vapour-sublimation techniques at a temperature of 2500°C. This produces small platelet crystals up to 1 cm in diameter and 0·1 cm thick.

Silicon carbide can be grown by the vapour-phase transport of silicon and carbon compounds in hydrogen. In this process, various sources of silicon and carbon, such as silane and the hydrocarbons, can be used.

Liquid-phase epitaxial processes pose considerable difficulties, as carbon has a low solubility in molten silicon. The usual technique adopted is to transport SiC from a source to a seed using a temperature gradient established in molten silicon. Alternatively, liquid-phase processes can be carried out using chromium as a solvent in a high vacuum, and some success has also been achieved by using the rare-earth elements in a similar fashion.

Silicon carbide can crystallize in either cubic (α-SiC) or hexagonal forms, and a large number of intermediate polytypes also exist. In describing these polytypes, it is convenient to consider the lattice to be made up of a stack of Si–C layers; each successive layer in the stack can be added in one of two orientations to give either the cubic (c) or hexagonal (h) structures. Alternatively, a mixed stacking order can be used, giving rise to the polytypes. These polytypes are designated by a number specifying the number of layers required to complete a sequence and a letter specifying the form of the primitive cell. For example, the polytype 6H has the six-layer stacking pattern hcchcc and an hexagonal structure.

The 6H polytype has an indirect band gap of 3·0 eV and shows a variety of radiative recombination processes, depending on the impurity levels involved. *n*-type material is obtained by doping with nitrogen, and deep acceptor levels can be introduced by using aluminium or boron. The transition energies that can be observed in such material are summarized in Table 4.3.

4.5.2 Silicon carbide devices

Typical SiC electroluminescent diodes are made by simultaneously diffusing boron and aluminium into nitrogen-doped material.[26] The diffusion

TABLE 4.3. Radiative transitions in silicon carbide (300 K).

Dopants	Photon energy (eV)	Photon wavelength (Å)
N	2·75	4500
N,Al	2·61, 2·51, 2·41	5000
N,B	2·07	6000

process requires a temperature in excess of 2000°C, and devices with an external efficiency of about 3×10^{-3} per cent. in the yellow region of the spectrum are obtained. Red- and green-emitting devices of a similar efficiency can also be made.[27]

Because of the inert chemical nature of SiC, devices made in this material are very stable. They can thus be operated at high temperatures and high current densities. Devices can be operated continuously at temperatures up to 400°C and current densities of up to 10 A/cm². However, under these conditions, their efficiency is considerably lower than at room temperature.

Although considerable effort has been devoted to the study of silicon carbide, this has not yielded practical devices with a useful efficiency. Thus it seems unlikely that silicon carbide devices will be competitive with those based on gallium phosphide or gallium arsenide–phosphide.

4.6 The II–VI compounds

Several of the II–VI compounds have a direct band gap that is greater than 2 eV. ZnS, for example, has a band gap of 3·6 eV and shows efficient radiative recombination when excited by an electron beam or ultraviolet radiation. However, since none of the materials that are of interest can be doped both *n* and *p* type, no electroluminescent *p–n* junction devices have been made from the binary compounds. Several ways in which efficient electroluminescent devices might be made have been explored, but up to now these have met with little success.

4.6.1 *Homojunction diodes*

CdTe is the only II–VI compound with an appreciable band gap which shows amphoteric conduction. Thus *p–n* junctions can be made in this material, but the observed efficiencies are low and the band gap (1·6 eV) does not give rise to visible radiation.

ZnSe and ZnTe can be doped *n* and *p* type, respectively, so that it should be possible to find a range of materials of the type $ZnSe_xTe_{1-x}$ with amphoteric behaviour. At the same time, the band gap of the material should lie in the range 2·6–2·2 eV. Aven[28] has shown that the material is amphoteric for $0.3 < x < 0.6$. This material can be grown by vapour-phase techniques, and *p–n* junctions can be formed by the diffusion of aluminium from a molten Zn–Al alloy at 100°C.

Yamento and Itoh[29] have shown that the materials $Cd_xMg_{1-x}Te$ have amphoteric properties over a range of compositions. Diodes were made by diffusing phosphorus into material with the composition $Cd_{0.65}Mg_{0.35}Te$. At room temperature these had an electroluminescent efficiency of 10^{-4} per cent. at 6800 Å.

4.6.2   Heterojunction diodes

In principle, it should be possible to produce a *p–n* junction between semiconductors with different band gaps. Thus it might be possible to obtain a junction between *n*-type ZnSe and *p*-type ZnTe. In such a situation, minority-carrier injection into the material with the lower band gap is to be expected, but, in general, this approach has been unsuccessful in producing electroluminescent diodes.

4.7   Comparison of materials

The experimental results that have been discussed in the previous sections are summarized in Table 4.4.

For diodes emitting radiation in the visible region of the spectrum, the basis for comparison is the overall luminous conversion efficiency of the device. This is obtained by multiplying the power efficiency of the device by the luminous efficiency of the radiation emitted. In a practical situation, this is not the only consideration when selecting a material, and factors such as the diode structure on which the results were obtained and economic factors must also be taken into account.

From the results given in Table 4.4, it is apparent that gallium phosphide devices have shown the highest visual efficiencies in the red and green regions of the spectrum. However, it must be remembered that the efficiencies quoted for this material apply to devices emitting radiation into a solid angle of 4π steradians, whereas many of the other results apply to devices which emit only into a hemisphere (2π steradians). Moreover, the epitaxial techniques used for the growth of *p–n* junctions are not entirely under control, and practical efficiencies for devices that can be readily obtained are a factor of 3–10 below those quoted.

The phosphor-coated gallium arsenide diode, emitting at wavelengths of 6600 and 5500 Å, would appear to be a significant competitor to gallium phosphide devices. However, the high efficiencies quoted are only obtained at impractically high current densities, where the thermal dissipation of the device and its reliability become problems. Thus this approach is of less practical significance than might be assumed.

TABLE 4.4. Comparison of electroluminescent diode efficiencies.

Material	Diode	Emission wavelength (Å)	Luminous efficiency of radiation (lm/W)	Laboratory results		Practical devices	
				Quantum efficiency (%)	Diode luminous efficiency (mlm/W)	Quantum efficiency (%)	Diode luminous efficiency (mlm/W)
GaAs	Zn diffused	9000		2		1	
	Si amphoteric	9300		28		10	
	Phosphor coated	6600	30	1	300		
		5500	600	0·1	600		
		4700	60	0·01	6		
$GaAs_xP_{1-x}$	Zn diffused	6500	50	0·2	100	0·1	50
$Ga_xAl_{1-x}As$	Zn diffused	6550	40	0·15	60		
GaP	Zn–O doped	6900	15	7	1000	1	150
	N doped	5650	600	0·5	3000	0·05	300
SiC	B doped	6000	300	$\sim 10^{-3}$	3		
	Al doped	5000	200	$\sim 10^{-3}$	2		

Although the luminous efficiency of gallium arsenide–phosphide is considerably lower than those of gallium phosphide devices, the advanced technology of the material yields consistent devices with an economic

FIGURE 4.20. A gallium arsenide–phosphide numerical display (Hewlett–Packard 7000 series). This device has a hybrid structure in which 27 light-emitting diodes and a silicon integrated circuit are mounted on a ceramic substrate. The diodes are ranged in a 7×5 matrix but, since the device is designed to display the numerals 0–9, only 27 diodes are required. The silicon circuit decodes a b.c.d. input and drives the required elements. At a drive current of 10 mA, each element has a luminance of 600 cd/m^2 (200 ft L). (Reproduced by permission of Hewlett–Packard Ltd.)

process. Thus gallium arsenide–phosphide has been exploited extensively, and a wide range of devices is available.

For devices emitting in the infrared region of the spectrum, no simple comparison of results is possible. The choice of a particular type of diode will depend on details of the application, including the detector to be used and the modulation frequency which is required. Gallium arsenide diodes that are amphoterically doped with silicon give the highest efficiencies, but can only be modulated at frequencies of 1–10 MHz. Zinc-diffused gallium arsenide diodes have an efficiency which is an order of magnitude lower than that of the silicon-doped device, but can be modulated at frequencies greater than 100 MHz.

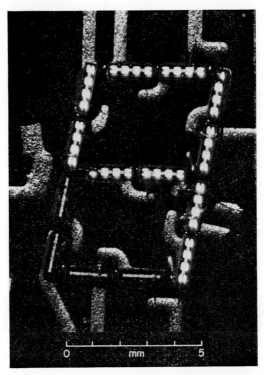

FIGURE 4.21. A seven-segment gallium arsenide–phosphide display (Monsanto MAN-1 series). This device has seven segments mounted on a ceramic substrate. Each of these segments consists of two diodes connected in series so that an operating voltage of 4 V is required. At a current of 20 mA, each element has a luminance of 1000 cd/m² (350 ft L)

4.8 Conclusions

Some of the wide range of electroluminescent display devices which have been developed are illustrated in Figures 4.20–4.24. The devices that are shown include both hybrid and monolithic structures made in gallium arsenide–phosphide or gallium phosphide. These examples are not intended to indicate the limitations of either of these technologies or materials, but to illustrate some of the ways in which they have been utilized in practical devices.

At present, most of the display devices that are available are based on gallium arsenide–phosphide and thus emit in the red spectral region. However, as the technology of other materials improves, it is to be expected that a similar range of devices, emitting in other regions of the spectrum, will become available.

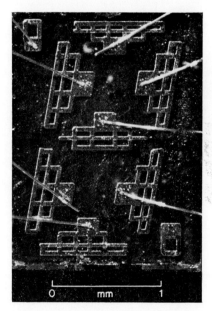

FIGURE 4.22. A seven-segment gallium phosphide numerical display (Ferranti type 2000 series). A monolithic device with seven segments diffused into nitrogen-doped GaP and emitting in the green spectral region. Each element has a luminance of 2000 cd/m^2 (650 ft L) at a drive current of 10 mA. (Reproduced by permission of Ferranti Ltd.)

FIGURE 4.23. A 7×5 gallium phosphide display
(S.E.R.L. experimental device). 35 GaP light-
emitting diodes are bonded to a ceramic substrate
and connected to give a coordinate-addressed
matrix. Red- or green-emitting diodes can be used
to give a luminance of 1000 cd/m² (350 ft L) at a
mean current of 10 mA per element. (Crown
Copyright, reproduced by permission of the
Controller, Her Majesty's Stationery Office.)

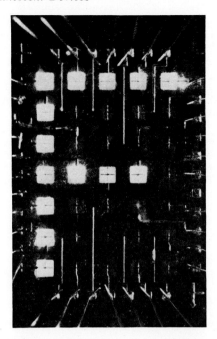

FIGURE 4.24. A 7×5 monolithic gallium phosphide display (Ferranti experimental device). 35 green-emitting diodes are diffused into a GaP substrate. Contacts are made by a conductor pattern on the surface of the semiconductor. (Reproduced by permission of Ferranti Ltd.)

4.9 References

1. H. Welker, *Z. Naturforsch.*, **7a**, 744 (1952).
2. J. I. Pankove and J. E. Berkeyheiser, *Proc. IRE.*, **50**, 1976 (1962).
3. K. L. Konnerth, J. C. Marinace and J. C. Topalian, *J. Appl. Phys.*, **41**, 2060 (1970).
4. H. Rupprecht, in *Gallium Arsenide, Proc. Intl. Symp., Reading, 1966*, IPPS, London, 1967, p. 57.
5. I. Ladany, *J. Appl. Phys.*, **42**, 654 (1971).
6. S. V. Galginaitis and G. E. Fenner, in *Gallium Arsenide, Proc. Intl. Symp., Dallas, 1968*, IPPS, London, 1969, p. 131.
7. F. W. Ostermayer, *Trans. Met. Soc. AIME.*, **2**, 747 (1971).
8. S. V. Galginaitis, *Trans, Met. Soc. AIME.*, **2**, 757 (1971).
9. J. E. Geusic, F. W. Ostermayer, H. M. Marcos, L. V. Van Uitert and J. P. van der Ziel, *J. Appl. Phys.* **42**, 1958 (1971).
10. C. J. Nuese, J. J. Tietjen, J. J. Gannon and H. F. Grossenberger, *J. Electrochem. Soc.*, **116**, 248 (1969).

11. M. Pilkuhn and H. Rupprecht, *J. Appl. Phys.*, **36**, 684 (1965).
12. H. P. Maruska and J. I. Pankove, *Solid State Electron.*, **10**, 917 (1967).
13. A. H. Herzog, W. O. Groves and M. G. Craford, *J. Appl. Phys.*, **40**, 1830 (1969).
14. W. O. Groves, A. H. Herzog and M. G. Craford, *Appl. Phys. Lett.*, **19**, 184 (1971).
15. E. G. Dierschke, L. E. Stone and R. W. Haisty, *Appl. Phys. Lett.*, **19**, 98 (1971).
16. H. Kressel, F. Z. Hawrylo and N. Almeleh, *J. Appl. Phys.*, **40**, 2248 (1969).
17. Quoted at AIME Conf. Recent Advances in Electronic Materials, San Fransisco, 1971.
18. J. Starkiewicz and J. W. Allen, *J. Phys. Chem. Solids*, **23**, 881 (1962).
19. S. J. Bass, P. E. Oliver and F. E. Birbeck, *J. Cryst. Growth.*, **2**, 196 (1968).
20. D. G. Thomas, *Brit. J. Phys. D* (*Appl. Phys.*), **2**, 637 (1969).
21. R. Nicklin, C. D. Mobsby, G. Lidgard and P. B. Hart, *J. Phys. C* (*G.B.*) **4**, L344 (1971).
22. R. H. Saul, J. Armstrong and W. H. Hackett, *Appl. Phys. Lett.*, **15**, 229 (1969).
23. R. A. Logan, H. G. White and W. Wiegman, *Solid State Electron.*, **14**, 55 (1971).
24. M. A. Carter, A. Mottram, A. R. Peaker and P. D. Sudlow, *Nature*, **231**, 469 (1971).
25. R. W. Brander, *Proc. IEE.*, **116**, 329 (1969).
26. R. M. Potter, J. M. Blank and A. Addamiano, *J. Appl. Phys.*, **40**, 2253 (1969).
27. A. A. Vasenkov, I. I. Kruglov, V. I. Pavlichenko, I. V. Ryzhikov and V. P. Sushkov, *IEEE. J. Solid-State Circuits*, **4**, 421 (1968).
28. M. Aven, *Appl. Phys. Lett.*, **7**, 146 (1965).
29. R. Yamamoto and K. Itoh, *Jap. J. Appl. Phys.*, **8**, 341 (1969).

5

p–n Junction Lasers

5.1 Introduction

A *p–n* junction laser consists essentially of a *p–n* junction that is constructed within a resonant optical cavity (Figure 5.1). When a current is passed through the device, in a forward direction, the junction region emits radiation. The optical cavity provides feedback so that stimulated emission can occur and, at a sufficiently high current, the device oscillates and laser action takes place. This critical current is known as the threshold current of the device. Various shapes of resonant cavity have been proposed and experimental devices made, but the Fabry–Pérot structure is the only one that is of any practical significance. In this structure, the device is in the form of a parallel-epiped with two opposite parallel surfaces forming the partially reflecting faces of the cavity. The junction plane is perpendicular to these two faces. Electrical contacts are applied to the two surfaces of the device that are parallel to the junction, and these surfaces must also be designed to allow the heat that is dissipated in the device to be removed. Since a laser operates at current densities of 10^3 A/cm^2 or more, these thermal problems can be severe.

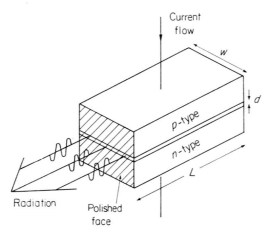

FIGURE 5.1. Schematic diagram of a semiconductor laser

113

TABLE 5.1. Semiconductor laser materials—*p-n* junction lasers.

Material		Reference	Photon energy (eV)	Wavelength (μm)
Gallium arsenide–phosphide	$GaAs_xP_{1-x}$	1–4	1·9–1·4	0·65–0·9
Gallium–aluminium arsenide	$Ga_xAl_{1-x}As$	5,6	1·9–1·4	0·65–0·9
Gallium arsenide	$GaAs$	7–9	1·4	0·9
Indium phosphide	InP	10,11	1·35	0·91
Gallium antimonide	$GaSb$	12	0·83	1·5
Indium arsenide–phosphide	$InAs_xP_{1-x}$	13	0·8	1·6
Gallium–indium arsenide	$Ga_xIn_{1-x}As$	14	0·7–0·6	1·8–2·1
Gallium–indium phosphide	$Ga_xIn_{1-x}P$	15	1·6	0·76
Indium arsenide	$InAs$	16,17	0·4	3·1
Indium arsenide–antimonide	$InAs_xSb_{1-x}$	18	0·39	3·2
Indium antimonide	$InSb$	19	0·23	5·4
Lead sulphide	PbS	20	0·29	4·3
Lead telluride	$PbTe$	21	0·19	6·5
Lead selenide	$PbSe$	22	0·14	8·5
Lead–tin telluride	$Pb_xSn_{1-x}Te$	23,24	0·2–0·05	6·5–30
Lead–tin selenide	$Pb_xSn_{1-x}Se$	25	0·1	10–12

The optical gain achieved in a semiconductor is high, with $10-100$ cm^{-1} being typical. Thus laser action can be achieved in short cavities and, in practical devices, the length of the cavity is usually between 10^{-1} and 10^{-2} cm.

5.2   Materials for *p–n* junction lasers

At present, only the direct band-gap semiconductors have been used as the basis of semiconductor lasers. Although there appears to be no fundamental reason why an indirect band-gap material should not be used in this way, estimates of the current densities that would be required suggest that this would be an unprofitable approach. On the other hand, all the direct-gap materials with a reasonably advanced technology have been made into lasers, and emission in any part of the spectrum from 0.3 μm to 30 μm can be demonstrated. A list of the materials in which laser action has been demonstrated is given in Tables 5.1 and 5.2.

Table 5.1 gives materials that have been used as *p–n* junction lasers. By far the major part of the experimental work in this subject has been concerned with devices made from gallium arsenide, and such devices are the only ones that are readily available. This is partly because the technology of gallium arsenide is considerably more advanced than that of any other relevant material, and also because the emission from gallium arsenide devices occurs at a wavelength (0.9 μm) that is well suited to detection with silicon solid-state detectors. Thus this chapter will be almost entirely concerned with gallium arsenide devices.

For some purposes, however, emission at wavelengths other than 0.9 μm is required, and then other semiconductors have been used. The most successful work has been based on the ternary III–V compounds such as $GaAs_xP_{1-x}$ or $Ga_xAl_{1-x}As$, giving rise to lasers emitting at wavelengths as short as 0.64 μm.

At the longer-wavelength end of the spectrum, diode laser action has been demonstrated in $Pb_xSn_{1-x}Se$ and $Pb_xSn_{1-x}Te$ at wavelengths out to 30 μm, but here again the devices have, as yet, found little practical application.

Table 5.2 lists materials in which laser action has been obtained using electron-beam or optical excitation. Some of these materials are also included in Table 5.1, and there is no basic reason why all the materials which have made junction lasers should not make optical or electron-beam-excited lasers. The other materials listed here are direct band-gap materials that are not amphoteric and can thus not make *p–n* junction lasers. These materials can only be excited optically or by an electron beam.

A wide range of materials have been excited in this way, and the available wavelength range, obtained using zinc sulphide and zinc oxide, extends into the ultraviolet region of the spectrum. However, neither electron beam nor

TABLE 5.2. Semiconductor laser materials—electron-beam or optically pumped lasers.

Material		Reference[a]		Photon energy (eV)	Wavelength (μm)
		e	o		
Zinc sulphide	ZnS	26		3·8	0·33
Zinc oxide	ZnO	27	28	3·3	0·38
Zinc selenide	ZnSe	29		2·7	0·46
Zinc telluride	ZnTe	30,31		2·3	0·53
Cadmium sulphide	CdS	32,33	34	2·5	0·50
Cadmium sulphide–selenide	CdS_xSe_{1-x}	35		2·5–1·8	0·5–0·7
Cadmium selenide	CdSe	36	37,38	1·8	0·69
Cadmium telluride	CdTe	39	40	1·6	0·78
Gallium nitride	GaN		41	3·45	0·36
Gallium selenide	GaSe	42		2·1	0·59
Gallium arsenide–phosphide	$GaAs_xP_{1-x}$	43		1·8	0·7
Gallium arsenide	GaAs	44	45	1·4	0·85
Gallium antimonide	GaSb	46		0·83	1·5
Indium arsenide	InAs	47	48	0·4	3·1
Indium antimonide	InSb	49	50	0·23	5·4
Cadmium silicon arsenide	$CdSiAs_2$	51		1·6	0·77
Cadmium tin phosphide	$CdSnP_2$	52		1·24	1·0
Cadmium phosphide	Cd_3P_2		53	0·6	2·1
Tellurium	Te	54		0·33	3·7
Mercury–cadmium telluride	$Hg_xCd_{1-x}Te$		55	0·3	3·8–4·1
Lead sulphide	PbS	56		0·29	4·3
Lead telluride	PbTe	56		0·19	6·5
Lead selenide	PbSe	56		0·14	8·5
Lead–tin telluride	$Pb_xSn_{1-x}Te$		57	0·08	15

[a] e = electron-beam pumped
o = optically pumped

optical excitation gives rise to a useful device and, for practical purposes, it will only be necessary to consider *p–n* junction devices in any detail.

5.3 The fabrication of laser structures

In discussing the fabrication of semiconductor lasers, it will be convenient first to consider the techniques used to form the *p–n* junction region in which spontaneous and stimulated emissions take place. The structure of this region is of fundamental importance to the operation of the laser, and must be designed to confine, as far as possible, the injected carriers and the optical flux to a narrow active region.

The incorporation of the junction into a resonant cavity to make a basic laser and the construction of practical device structures will also be discussed.

5.3.1 The p–n junction

The progressive development of gallium arsenide lasers can be clearly identified with the adoption of several distinct techniques for making the *p–n* junction. Although the first devices were based on diffused junctions, these have now been almost entirely supplanted by epitaxial techniques, which have been developed so that complex multilayer structures comprising materials of different conductivity type, doping level and band gap can be grown under very close control.

[i] *Diffused junctions.* The first junction lasers were made by diffusion techniques. With gallium arsenide, the procedure adopted was to diffuse zinc into *n*-type material to give a *p–n* junction 5–50 μm below the surface of the semiconductor. Although this diffusion process has been studied in considerable detail, the exact conditions needed to give the best lasers were usually derived empirically. For example, suitable junctions can be formed by diffusing zinc into gallium arsenide from a zinc–arsenic vapour at a temperature of 850°C and then annealing the diffused semiconductor at a lower temperature.[58] Lasers made in this way can have threshold current densities of 500 A/cm^2 at 77 K and 4×10^4 A/cm^{-2} at 300 K. The diffusion technique is very versatile, and has been used to produce lasers from a wide range of semiconductors.

[ii] *Epitaxial junctions.* Theoretical studies have suggested that a junction laser should be made in heavily doped material to obtain a low threshold current. Such junctions can best be made using the liquid-phase epitaxial process.[59] For example, *p*-type gallium arsenide can be grown onto an *n*-type substrate from a zinc-doped solution of gallium arsenide in gallium.

The structure of a typical device is shown in Figure 5.2. With this structure, small variations of refractive index on either side of the active region help to confine the radiation to this active region. However, there is still a considerable penetration of the optical field into absorbing regions on either side of the active region.

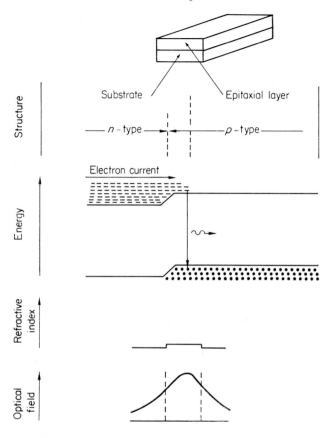

FIGURE 5.2. The structure of a *p–n* junction laser, showing the band structure, refractive index and optical-field distribution in the junction region

Lasers made by the epitaxial technique described here show a considerable improvement over diffused devices and have threshold current densities of about 2×10^4 A/cm^2 at room temperature.[60]

[iii] *Heterojunction structures*. A considerable reduction in the room-temperature threshold current of a gallium arsenide laser can be achieved by growing the heterojunction structure shown in Figure 5.3.[61–64] This structure can be obtained by growing a *p*-type zinc-doped gallium–aluminium arsenide layer onto an *n*-type GaAs substrate. During the growth of this layer, from a liquid-phase gallium solution, zinc diffuses from the grown layer into the substrate so that the *p–n* junction is formed within the gallium arsenide. This structure has two merits:

(a) The discontinuity in the refractive index that occurs at the GaAs–Ga$_x$Al$_{1-x}$As interface partially confines the optical flux to the active region of the device and thus lowers the optical losses of the cavity.

(b) Since Ga$_x$Al$_{1-x}$As has a wider band gap than GaAs, a potential barrier occurs at the GaAs–Ga$_x$Al$_{1-x}$As interface. This potential barrier confines the electrons that are injected across the forward-biased *p–n* junction to a narrow region, and thus increases the gain in the device.

Lasers made with this heterojunction structure (which is sometimes known as the close-confinement structure) have a room-temperature threshold current density of about 10^4 A/cm^2.

A simple extension of the techniques used to grow heterojunction devices can be used to make lasers in which the active region has a composition Ga$_x$Al$_{1-x}$As instead of GaAs.[65] These have similar threshold properties, but emit at a shorter wavelength. For example, using suitable compositions,

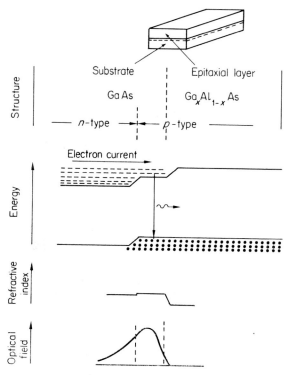

FIGURE 5.3. The structure of a heterojunction laser, showing the band structure, refractive index and optical-field distribution in the junction region

E

efficient laser action at wavelengths shorter than 8000 Å (at 300 K) has been
reported.

[iv] *Double-heterojunction structures.* A further development arising from
the epitaxial techniques for growing laser structures is the double-heterojunction device shown in Figure 5.4.[66-69] In this structure, a narrow active
region of GaAs is bounded on both sides by regions of $Ga_xAl_{1-x}As$, which
effectively confine the injected electrons and the optical flux to the active
region, which is about 1 μm wide.

Structures of this complexity can be grown by an epitaxial technique
(described in Section 3.2.1) in which several layers are grown in succession

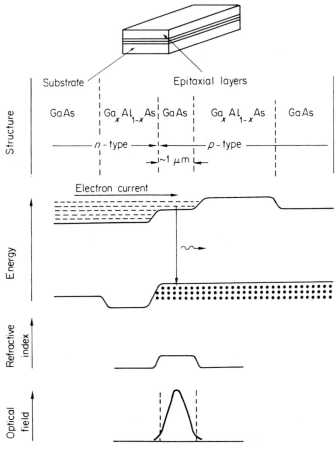

FIGURE 5.4. The structure of a double-heterojunction laser,
showing the band structure, refractive index and optical-field
distribution in the junction region

using a crucible containing several melts of different composition. In this way, the laser structure does not have to be removed from the furnace at each stage, and temperature-control problems are minimized.

The double-heterojunction structure has three merits:

(a) The discontinuity of refractive index at the boundaries of the active region gives a waveguide effect that minimizes the penetration of the optical flux into the absorbing material on either side of the active region of the device.

(b) The wider band gap of the $Ga_xAl_{1-x}As$ gives a potential barrier at the boundary of the active region, and thus confines the injected minority carriers to this region.

(c) The *p–n* heterojunction favours the injection of electrons into the *p*-type active region, and minimizes the unwanted injection of holes across the junction into the *n*-type region.

As a result of the improvements brought about by the double-heterojunction structure, lasers with threshold current densities as low as 940 A/cm^2 at room temperature have been produced.[70]

The growth techniques used to fabricate double-heterojunction structures usually result in a small proportion of aluminium being incorporated in the GaAs active region. Because of this, the emission wavelength of these devices is about 200 Å shorter (i.e. 8800 Å at room temperature) than that of other GaAs devices. The wavelength can be made still shorter by intentionally adding a higher proportion of aluminium to the active region. In this way, devices emitting at 7500 Å with threshold current densities of 3×10^3 A/cm^2 at room temperature have been made. Other devices, emitting at 7000 Å, have operated continuously at room temperature.[71]

In the double-heterojunction structure both the injected carriers and the optical flux are confined to the same narrow active region; this has some disadvantages when the active region is made very narrow. For example, the beam divergence of the emitted radiation can be quite large and the high optical flux density can damage the cavity at modest output powers.

[v] *Four- and five-layer structures.* Some of the disadvantages of the double-heterojunction lasers might be overcome by devising devices in which the injected carriers and the optical flux are confined to different regions of the structure. For example, carriers might be confined to a narrow GaAs region by a GaAs–$Ga_xAl_{1-x}As$ discontinuity whose parameters would be chosen so as not to confine the optical flux. The optical flux would then occupy a larger region that would also be bounded by discontinuities in the structure, and designed not to absorb radiation.

Work along these lines has been described by Kressel, Lockwood and Hawrylo.[72] Their structure—which they have called the large-optical-cavity (l.o.c.) structure—consists essentially of four layers with three discontinuities. Carriers are confined to a narrow, p-type GaAs region in which gain occurs. The optical flux is confined to a low-loss n-type GaAs region by a GaAs–$Ga_xAl_{1-x}As$ boundary. The structure is made by a number of epitaxial growths on a gallium arsenide substrate and has given threshold current densities of 2500 A/cm^2.

The development of symmetrical structures, in which a GaAs active region is in the centre of a $Ga_xAl_{1-x}As$ optical waveguide region, has been discussed by Thompson.[73] These structures demand extremely close control of the epitaxial growth process in order to be successful but may lead the way to further improvements in gallium arsenide lasers.

5.3.2 The laser cavity

The small cavities required for junction lasers are usually constructed by taking advantage of the natural cleavage planes of the semiconductor. Gallium arsenide has a cubic structure and cleaves along the [110] planes. Thus a p–n junction that is grown on a [100] plane will be perpendicular to one of the natural cleavage planes of the crystal. To make use of this property, epitaxial junctions are grown on a substrate cut on the [100] plane, and the slice is then polished to reduce its total thickness to about 100 μm. At this point, electrical contacts are put on the surfaces of the slice, which can then be cleaved into bars. The width of these bars is made equal to the required cavity length. The bars are then sawn into dice of the required size. The rough sawn edges of the laser die help to inhibit any tendency that the laser might have to oscillate in modes perpendicular to the desired direction in the Fabry–Pérot cavity.

Gallium arsenide has a refractive index of 3·5, and a GaAs–air interface has a reflectivity of 30 per cent. Thus the cleaved faces of a gallium arsenide Fabry–Pérot structure can be used as the cavity mirrors. However, in order that all the radiation from a device should emerge from one end of the cavity, the reflectivity of one face must be increased to a value approaching 100 per cent. This can readily be done by first coating the face with a layer of an insulator such as silicon monoxide, and then coating with a layer of a good reflector such as silver or aluminium.

5.3.3 Practical laser structures

The small physical size of semiconductor lasers and the high power densities at which they are required to operate pose some interesting problems in the design of practical devices. The structure adopted must provide an adequate heat sink, and must also allow the device to be used in an optical system, which may involve focusing the radiation with a lens or coupling the radiant energy into a fibre-optic system. To obtain the most effective use of semiconductor lasers one would, ideally, design a device for a particular requirement. This approach is not usually feasible, and most systems must use devices that are designed to meet a range of requirements. The design of some typical devices will be described in this section. Although, at this stage, it will be convenient to outline the applications for which these devices are intended, a detailed description of their properties will be reserved for subsequent sections of this chapter, and a detailed discussion of their applications will be found in Chapter 6.

[i] *Single lasers.* One of the most simple and practical device structures is shown in Figure 5.5. In this structure, one side of the laser die is bonded to a heat sink, while connection is made to the other side by means of a coaxial wire which passes through the stud. The device is designed to be screwed into a larger heat sink, which may be at room temperature or be cryogenically cooled, but, in general, devices used in this way can only be operated at quite low duty cycles, typically 1–10 per cent. at 77 K or 0·1–1 per cent. at 300 K.

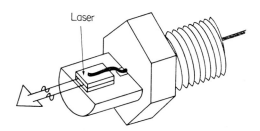

Laser

FIGURE 5.5. A device structure with a laser mounted on a hexagonal stud

[ii] *High-mean-power devices.* To obtain the maximum mean power from a laser, it is necessary to remove heat from both sides of the laser die, and several structures designed to achieve this have been developed. Although these structures have been successful technically, they have not found wide application, since they are difficult and expensive to produce.

In the device shown in Figure 5.6, the laser is bonded between two tungsten discs that are separated by a slice of a suitable insulator.[74] The tungsten discs have a high thermal conductivity at low temperatures, and also have a thermal-expansion coefficient similar to that of the laser. The insulating spacer gives the structure mechanical strength, and also provides a thermal path so that both sides of the laser are cooled when one side of the structure is mounted on a heat sink.

[iii] *High-duty-cycle devices.* One of the thermal limitations of a device structure occurs at the interface between the semiconductor die and the heat

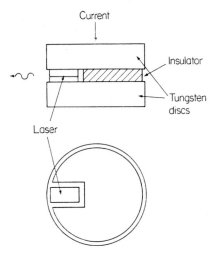

FIGURE 5.6. A laser structure designed
for high power dissipation

sink. For a device of a given area, the thermal resistance of this interface
is made as small as possible if the device has a length that is much greater
than its width. Thus, to obtain structures that will operate at high duty
cycles, the laser die should be long and narrow and mounted on a heat sink
of high thermal conductivity.

A structure that has been particularly successful in achieving operation
at high duty cycles is illustrated in Figure 5.7.[75] In the simplest form of
this structure, a narrow *p–n* junction is diffused through a silica mask and
a contact applied over this mask. The heat sink is usually made of copper,
but an improved performance is obtained if the heat sink is made of type II
diamond to which the diffused surface is bonded.

This structure, which is commonly called the 'stripe-geometry' structure,

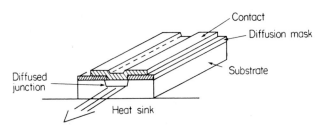

FIGURE 5.7. Stripe-geometry laser for high-duty-cycle
operation

can be adapted to use epitaxial junctions, and devices made in this way have operated continuously at room temperature.

[iv] *Multiple-laser devices.* Because semiconductor lasers operate at a high current and low voltage, there is an advantage to be gained if several devices are operated in series. One way in which this can be achieved is by combining several laser dice in a stack, as shown in Figure 5.8. This structure has poor thermal properties, and thus cannot operate at high duty cycles, but high peak powers can be obtained. To achieve the highest average radiance, the thickness of each die must be as small as possible, so as to keep the junctions close together, but, even with the best structures, the radiance that can be achieved is an order of magnitude lower than that of

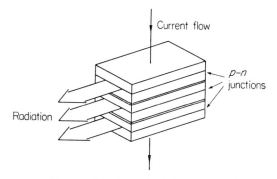

FIGURE 5.8. A stack of three lasers

a single laser junction. However, since, in many situations, the full radiance of the linear laser source cannot be utilized, the laser stack is not necessarily a disadvantage.

5.4   The threshold properties of a junction laser

5.4.1   General observations

It is instructive to observe the appearance of a *p–n* junction laser as current through it is increased; if the laser emits in the near infrared, these observations can easily be made with the aid of an image converter. At low current densities, the junction appears as a rather diffuse line that is emitting spontaneous radiation. As the threshold current is reached, the luminance of small regions of the junction increases markedly, and these regions occupy an increasing fraction of the junction as the current is further increased (Figure 5.9).[76] The two ends of the cavity have a similar, mirror image, appearance and the bright regions of the junction indicate filaments in which laser action is occurring. In the most efficient devices, essentially

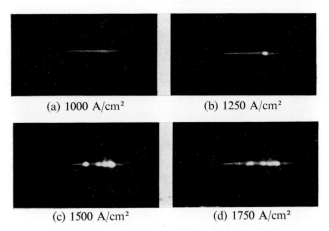

(a) 1000 A/cm²   (b) 1250 A/cm²

(c) 1500 A/cm²   (d) 1750 A/cm²

FIGURE 5.9. The appearance of a laser junction at successively
higher current densities; the onset of laser action can be
seen in (b) (Crown Copyright, reproduced by permission of
the Controller, Her Majesty's Stationery Office)

the whole of the junction achieves the threshold condition over a very small
increase in current, and it is difficult to resolve filamentary structure in the
device.

The power emitted by a laser, as a function of the current that is applied
to it, is shown in Figure 5.10. Below threshold the output is small, but
increases rapidly above a critical value of current. The critical current
determined from this curve forms a convenient way of experimentally
measuring the threshold current of the device, and the value obtained in

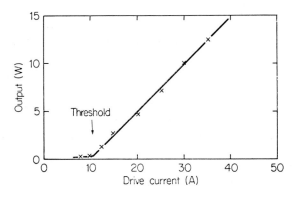

FIGURE 5.10. The output power from a laser as a
function of current (junction area $= 10^{-3}$ cm², 300 K)

this way agrees well with the current at which bright spots on the junction are first observed.

5.4.2 *The temperature dependence of threshold*

The threshold current of a junction laser increases rapidly with temperature, and much of the research on lasers has had the aim of reducing this temperature dependence to obtain lower threshold currents at room temperature. The temperature dependence of the threshold for some typical gallium arsenide lasers is shown in Figure 5.11. These curves represent the successive stages of development described in Section 5.2, and show the progress made in the development of semiconductor lasers since they were first made in 1962.

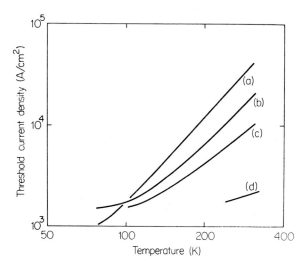

FIGURE 5.11. The variation of threshold current density with temperature for GaAs laser structures: (a) diffused, (b) epitaxial, (c) heterojunction and (d) double-heterojunction structures

A complete analytical description of the experimentally determined variation of threshold with temperature is not feasible, but the general features of the dependence are in fair agreement with theory. For diffused devices, the threshold increases slowly with temperature at low temperatures, but increases approximately as T^3 above 100 K. Epitaxial devices, particularly the heterojunction structures, show a smaller temperature dependence of threshold, and this can often be approximated by an exponential function of the form $I_0 \sim \exp T/T_0$.

5.4.3   Threshold parameters

At the threshold point of a laser, the gain and loss in the system are equal, and the following relationship holds:

$$g(J) = \alpha + \frac{1}{L}\ln\left(\frac{1}{R_1 R_2}\right) \tag{5.1}$$

where $g(J)$ is the gain per unit length of the cavity and is a function of the current density J, α represents the cavity losses (excluding power transmitted through the end faces of the device) and R_1 and R_2 are the reflectivities of the cavity faces.

The validity of this equation has been extensively studied by measuring the variation of threshold current with cavity length and reflectivity.[77,78] These measurements are difficult to make because of the problem of obtaining a reproducible and consistent series of devices, but have proved valuable in the investigation of the properties of different structures.

Some typical results, showing the threshold current density J_0 as a function of the reciprocal cavity length $1/L$, are shown in Figure 5.12. Comparing these results with the form of equation (5.1) indicates that the gain in the cavity is a linear function of the current density and can be expressed as:

$$g(J) = \beta J. \tag{5.2}$$

Although this relationship holds for a wide range of lasers, it is not universally valid, and must be used with caution. For example, in a number of cases the gain in a laser is a superlinear function of the current density, and equation (5.2) becomes

$$g(J) = \beta J^n \tag{5.2a}$$

where the experimental value of n lies between 2 and 3, depending on the laser structure.

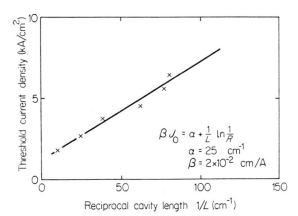

FIGURE 5.12. The variation of laser threshold current density with cavity length

The threshold parameters for some typical gallium arsenide devices made by the different techniques that were described in Section 5.2 are given in Table 5.3. This table also gives the threshold current density of a typical device with a cavity 0·05 cm long and silvered at one end, and the threshold current of a device made with these cavity parameters and a width of 0·02 cm.

At 77 K, the techniques used all give lasers with comparable threshold current densities of about 1 kA/cm². Although there is some variation among the quoted results, these are not significant when compared with the range of results obtained by different workers using similar techniques. This is not the case with results obtained at room temperature.

At room temperature, the gain coefficient in a diffused laser is about 10^{-3} cm/A and threshold current densities of 50 kA/cm² are typical. The use of epitaxial techniques to grow the junction gives devices with an improved gain, but also give an increase in the cavity loss α. The net result is that epitaxial devices have threshold current densities of about 30 kA/cm².

The single-heterojunction structure reduces the cavity losses and gives devices with a threshold current of about 12 kA/cm². The exact value of the threshold current depends on the width of the active region of the device, and is lowest when this width is about 2 μm (Figure 5.13).

In the double-heterojunction structure, the cavity losses are further reduced, and threshold current densities as low as 1 kA/cm² at room temperature have been reported.[70] In these devices, the gain is not a linear function of current density, but increases superlinearly, and it is thus not possible to specify a value for the gain coefficient β.

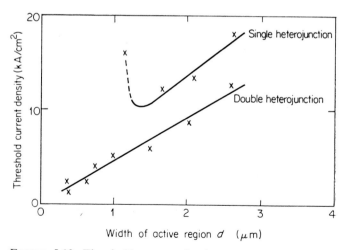

FIGURE 5.13. Threshold current density as a function of the width of the active region *d* for single- and double-heterojunction lasers at 300 K (cavity length = 250 μm)

TABLE 5.3. Typical threshold parameters of gallium arsenide laser structures.

Laser structure	Temperature (K)	Loss coefficient α (cm^{-1})	Gain coefficient β (cm/A)	Threshold current density[a] (A/cm^2)	Threshold current[b] (A)
Diffused	77	15	40×10^{-3}	1.0×10^3	1
Epitaxial	77	14	35×10^{-3}	1.1×10^3	1
Single heterojunction	77	10	40×10^{-3}	0.9×10^3	1
Diffused	300	24	1×10^{-3}	50×10^3	50
Epitaxial	300	90	4×10^{-3}	30×10^3	30
Single heterojunction	300	35	5×10^{-3}	12×10^3	12
Double heterojunction	300	20		2×10^3	2

[a] Threshold current densities are calculated for a cavity length of 0·05 cm.
[b] Threshold currents are for a device with a cavity length of 0·05 cm and a width of 0·02 cm.

The threshold current of double-heterojunction devices is also a function of the active region width *d*, as shown in Figure 5.13. The threshold current density decreases linearly with *d* as far as experimental results are available, giving a value of about 1 kA/cm^2 when $d = 0.2$ μm.[79]

These results are in good agreement with a theoretical analysis of the waveguide properties of the double-heterojunction structure, which takes into account the variation of the complex relative permittivity in the device.[80]

5.5 Efficiency

Before considering the efficiency of semiconductor lasers, it is first necessary to take note of the various ways in which this parameter is commonly specified.

5.5.1 Basic definitions

[i] *Quantum efficiency* η_q. The quantum efficiency of an electroluminescent device can be defined as the ratio of the number of photons emitted by the device to the number of electrons which flow through the device. If *P* is the power emitted at a drive current *I* and the photon energy is *hv*,

$$\eta_q = \frac{P/hv}{I/e}. \tag{5.3}$$

It is also sometimes of interest to consider the internal quantum efficiency of a device. This relates to the number of photons produced at the junction rather than those emitted from the device, and thus ignores the photons that are absorbed after emission. If this distinction needs to be made, the term 'external quantum efficiency' is used to specify the parameter which relates to the device output.

In a forward-biased *p–n* junction, the current which flows is an exponential function of the bias voltage *V*. At high current levels this bias voltage is essentially independent of the current, and is approximately equal to the band gap of the semiconductor. The bias voltage is thus equivalent to the photon energy of the emitted radiation, and equation (5.3) becomes, with sufficient accuracy:

$$\eta_q \approx \frac{P}{IV}. \tag{5.4}$$

[ii] *Incremental efficiency* $\Delta\eta$. From the form of the output-against-current curve for a laser (Figure 5.10), it can be seen that the efficiency of a device, as defined by equation (5.3), depends on the drive current. In specifying the properties of a device this is inconvenient, and it is thus common practice to define the incremental efficiency of a device to specify the slope of the output characteristic above threshold (Figure 5.10). Thus

$$\Delta\eta = \frac{P}{(I-I_0)V}. \tag{5.5}$$

[iii] *Power efficiency* η_P. In the discussion of efficiency, the power that is dissipated in the series resistance has, so far, been neglected. In a practical system, however, it is important and must be taken into account. If the series resistance of the laser is R, the total power dissipated in the device is

$$W = IV + I^2R \qquad (5.6)$$

and the power efficiency is

$$\eta_P = \frac{(I - I_0)V\Delta\eta}{IV + I^2R}. \qquad (5.7)$$

The power dissipated in the series resistance of a device becomes significant at high current densities, and is thus important at room temperature. For example, a device with a threshold current of 10 A, a series resistance of 0·2 Ω and an incremental efficiency of 25 per cent. would have a power efficiency of only 3 per cent. at a drive current of 30 A.

5.5.2   The efficiency of devices

Above the threshold condition, the internal quantum efficiency of a semiconductor laser is high. Estimates suggest that a value approaching 100 per cent. is possible, but, because of optical absorption, the external efficiency of a device is lower than this. Biard, Carr and Reed[81] have given the following formula for the incremental quantum efficiency of a laser:

$$\Delta\eta = \frac{\ln(1/R)}{\alpha L + \ln(1/R)}. \qquad (5.8)$$

This formula applies to a device in which both ends of the cavity have a reflectivity R. In the more usual situation, one end of the cavity has a reflectivity of 100 per cent., and equation (5.8) then becomes

$$\Delta\eta = \frac{\ln(1/R)}{2\alpha L + \ln(1/R)}. \qquad (5.9)$$

Equation (5.9) predicts the efficiency of a device that is perfectly uniform, and it is to be expected that the measured efficiency of practical devices will be somewhat lower than this. Bearing this in mind, the efficiency of devices is in fair agreement with values calculated from equation (5.9) with an appropriate value of α. This is well illustrated by the reported efficiencies of gallium arsenide lasers at room temperature and the values of α determined for the relevant structures. For example, Figure 5.14 shows the efficiency calculated from equation (5.9), taking $\alpha = 100$ cm^{-1} and 30 cm^{-1} with an internal quantum efficiency of 60 per cent. Also shown on this figure are some reported efficiencies of devices made with a simple epitaxial structure ($\alpha \approx 100$ cm^{-1}) and a single heterojunction structure ($\alpha \approx 30$ cm^{-1}). It can be seen that the heterojunction devices have an efficiency which is about a factor of two higher than that of simple epitaxial devices, and that incremental efficiencies of 40 per cent. can be achieved.

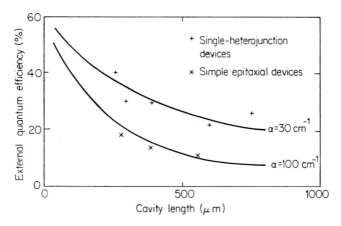

FIGURE 5.14. The efficiency of GaAs lasers as a function of
cavity length at room temperature

5.5.3 Radiation patterns

In considering the pattern of radiation emitted by a junction laser, it is common to discuss both near-field and far-field radiation patterns. The near-field pattern relates to the intensity of radiation close to the junction and is, essentially, the distribution of power across the junction, as shown in Figure 5.9. The far-field pattern describes the intensity of radiation at large distances from the laser, and can be observed by plotting the distribution of the intensity of radiation with a small-area detector or by allowing the radiation to fall on an infrared-sensitive film.

One of the important features of a gas laser is the very small divergence of the beam radiation that it produces. A semiconductor laser does not possess this feature and, in most practical systems, the radiation must be collimated with a lens or mirror. The source of radiation is the line formed by the intersection of the junction plane with the cavity face. One dimension of the source is equal to the width of the laser die, typically 100–1000 μm, the other dimension is given by the width of the active region of the device and is about 2 μm. If the source were coherent over the whole of this emitting area, the radiation pattern could be determined by considering the diffraction of radiation from this source. For example, the diffraction-limited beam from an aperture 2 μm $\times$ 100 μm would have a divergence of approximately 25° $\times$ 0·5°. In practice, devices are not coherent over the length of the junction, and radiation patterns with considerable divergence in both planes are observed.

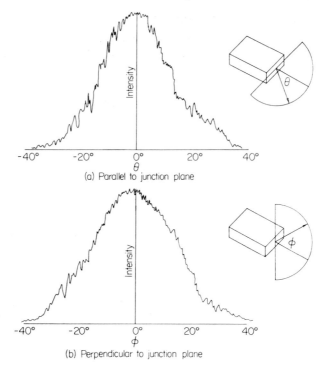

(a) Parallel to junction plane

(b) Perpendicular to junction plane

FIGURE 5.15. The far-field radiation pattern from a GaAs laser

A typical example of the far-field radiation pattern of a laser is shown in Figure 5.15. From this figure, it is apparent that the radiation has a divergence, to the half-intensity points, of about $\pm 15°$ in the junction plane and $\pm 20°$ in the plane perpendicular to the junction. Although the radiation patterns vary widely between different devices, in practical terms an optical system with an aperture of at least $f/2$ should be used to collimate the radiation from a semiconductor laser.

The radiation pattern from double-heterojunction devices is more complex than that from other devices and varies with the width of the active region d. For large values of d, complex optical modes are established in the cavity, giving a wide beam divergence with a number of lobes in the radiation. At the other extreme, the diffraction-limited beam from a narrow cavity also has a wide divergence.

The beam divergence of double-heterojunction lasers often exceeds $\pm 30°$ in the plane perpendicular to the junction. This imposes severe limitations on the optical system used with the device, and is a considerable disadvantage of this structure.

5.6   Thermal properties and operating conditions

Semiconductor lasers operate at high current densities, and the dissipated power density in a device is correspondingly high. Thus devices are usually operated under pulsed conditions to limit the dissipation to an acceptable level. However, continuous operation of carefully designed devices is possible, and has been achieved, even at temperatures somewhat above room temperature.

5.6.1   Continuous operation

It is convenient to discuss the continuous operation of a device by reference to Figure 5.16. In this figure, curve (a) represents the variation of the

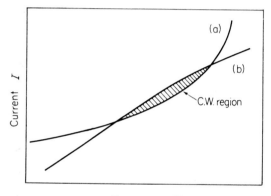

Temperature *T*

FIGURE 5.16. The conditions for continuous operation of a laser: (a) variation of threshold current with temperature and (b) variation of junction temperature with operating current

threshold current I_0 of a device with temperature T. The exact form of this curve depends on the structure of the device, but, in general, the threshold current increases rapidly with temperature, as a cubic or exponential function. Curve (b) represents the equilibrium temperature of the device when a current I passes through it. At low currents this will be a linear variation, as the device dissipation is proportional to current, but will tend to a square-law dependence as Joule heating becomes important at higher currents.

Continuous operation of a device is possible only in the region over which these two curves overlap, if such a region exists. It is in this region that the threshold current of the device, at the equilibrium temperature reached by the device, is lower than the current at which the device is operating.

The performance of a typical device capable of c.w. operation is shown in

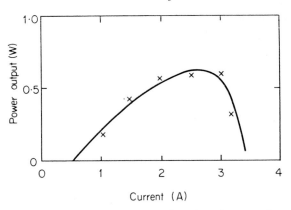

FIGURE 5.17. Power output as a function of current for
a c.w. laser (77 K)

Figure 5.17, and it can be seen that continuous operation occurs only over
a restricted range of currents.

The continuous operation of devices at cryogenic temperatures has been
reported from several laboratories. Output powers of 7 W at 4·2 K and 3 W
at 20 K have been obtained,[74,82] and at 77 K a power output of 1 W or
more is feasible. To achieve continuous operation at higher temperatures,
Dyment and D'Asaro have mounted very narrow devices on diamond heat
sinks and thus achieved c.w. operation above 200 K.[83] For some years
this remained the highest temperature for c.w. operation. However, the
lower threshold currents which have been made possible by the double-
heterojunction structure have resulted in devices that can be operated
continuously at room temperature with a power output of 20 mW.[84]

5.6.2 Pulsed operation

The majority of gallium arsenide lasers that are readily available must be
operated with a pulsed current. In fact, in most applications this mode of
operation is desirable, and the ease with which a semiconductor laser can
be pulsed is one of the attractions of the device.

To achieve laser action, the pulse of current that is applied to a device
must have a fast rise time, so that the threshold condition is reached before
the temperature of the device rises to such an extent that laser action at this
current is impossible. This is a complex problem to discuss quantitatively,
but in practice it is common to use pulse lengths of 1–10 μs at 77 K and
0·1–1 μs at room temperature. During the pulse, the temperature of the
device junction will rise, and this will be manifested as changes in the
spectrum and a decrease in the output power. However, for short pulses of
the duration suggested above, these effects will be small.

The performance of a typical device operating in a pulsed mode is given in Figure 5.18. This device has a threshold current of 10 A at 300 K and an incremental efficiency of 20 per cent. The peak and mean power outputs that are obtained when the laser is operated with 20, 30 or 40 A pulses over a range of duty cycles is shown.

The general form of the performance curves in Figure 5.18 can be predicted, from a knowledge of some of the basic properties, as follows.

With the previously defined symbols, the power P dissipated in the device and the junction temperature rise ΔT above ambient are given by

$$P = IV + I^2R \tag{5.10}$$

and
$$\Delta T = P\Theta\delta \tag{5.11}$$

where Θ is the thermal impedance of the device and δ is the duty cycle at which the drive current I is applied. Under these conditions, the operating

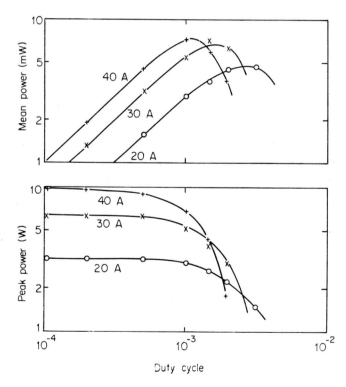

FIGURE 5.18. Peak and mean power outputs of a pulsed laser as a function of duty cycle (threshold current = 10 A, drive currents = 20, 30 and 40 A)

threshold current I_0' of the device can be calculated, and the device output is given by

$$\text{Peak power} = (I - I_0')V\Delta\eta \tag{5.12}$$

and $\qquad \text{Mean power} = (I - I_0')V\Delta\eta\delta. \tag{5.13}$

The peak power that can be obtained from a device is limited by the power density at which damage to the faces of the laser cavity occurs.[85,86] This power density varies with the junction structure of the device and the pulse length. For typical devices and pulse lengths, damage occurs when the power emitted from the device exceeds 50 W per millimetre of junction width. This corresponds to a power density in the junction region of about 10^6 W/cm^2.

For double-heterojunction lasers, the damage threshold is considerably lower than the value given above because the output flux is confined to a very narrow region which may be only 0.5 μm wide. This is about a factor of five smaller than the width of the active region in other structures and, as a result, the damage threshold is only about 10 W per millimetre of junction width. This, combined with the wide radiation pattern of double-heterojunction devices, gives the devices a much lower radiance than that obtained with homojunction or single-heterojunction structures.

5.7   Spectral properties

The emission from a semiconductor laser occurs at a wavelength that corresponds closely to the band-gap energy of the semiconductor, and the list of semiconductors given in Table 5.1 shows that the spectral region 0.6–30 μm can be covered by a suitable choice of material. This section will consider some of the details of the laser spectra, paying particular attention to gallium arsenide lasers.

At low currents the spectrum is that of the spontaneous emission from the junction, and has a half-intensity half width of approximately 100 Å. Figure 5.19 shows the photon energy and wavelength of the emission from GaAs diodes as a function of temperature. The emission has a photon energy about 70 meV less than the band-gap energy of pure GaAs, and shows some scatter owing to the variation of the doping level used in different devices.

As the current approaches the threshold value, the spectrum shows a mode structure and, at threshold, the power in one or more of these modes near the spectral peak increases rapidly (Figure 5.20).

At currents well above threshold, essentially all the radiated power occurs in a few modes extending over a wavelength range of approximately 25 Å (Figure 5.21). The observed modes correspond to the resonant longitudinal modes of the laser cavity, and their wavelength is given by

$$m\lambda = 2\mu L \tag{5.14}$$

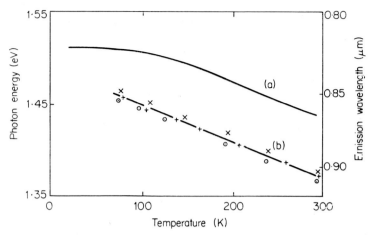

FIGURE 5.19. Emission wavelength and photon energy of GaAs lasers: (a) band gap of GaAs and (b) laser emission, showing variation of wavelength among devices

where m is an integer. The spacing between adjacent modes $\Delta\lambda$ is obtained by taking successive values of m in equation (5.14), bearing in mind that the refractive index is a function of wavelength. Thus, as in Section 2.5.3,

$$\Delta\lambda = \frac{\lambda^2}{2\mu L\left(1 - \frac{\lambda}{\mu}\frac{\partial\mu}{\partial\lambda}\right)}. \tag{5.15}$$

The dispersion term makes a significant contribution to this expression and, for a typical laser cavity length of 400 μm, the mode spacing is 2·5 Å. If the spectrum is observed with greater resolution, additional structure, due to the other resonant modes of the laser cavity, can be observed.

When the temperature of a laser is changed, the laser spectrum changes owing to two distinct effects.

Since the band gap of a semiconductor decreases with increased temperature, the wavelength of the spontaneous emission of a diode and the mean wavelength of the laser emission increase with temperature. In gallium arsenide the emission wavelength shifts from 8500 Å at 77 K to 9000 Å at 300 K, corresponding to a temperature coefficient of approximately 2·5 Å/K.

The second effect is owing to the temperature dependence of the optical properties of the semiconductor. It can be shown that the wavelength of a spectral mode has a temperature coefficient which is given by

$$\frac{1}{\lambda}\frac{d\lambda}{dT} = \frac{1}{\mu}\frac{\partial\mu}{\partial T}\left(1 - \frac{\lambda}{\mu}\frac{\partial\mu}{\partial\lambda}\right)^{-1}. \tag{5.16}$$

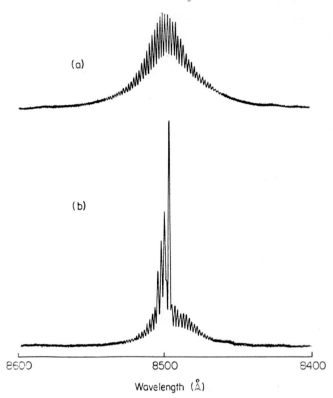

FIGURE 5.20. Laser spectra near the threshold: (a) below threshold and (b) above threshold

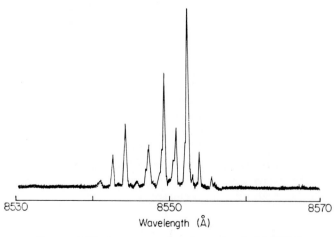

FIGURE 5.21. Laser spectrum above threshold (77 K)

For a gallium arsenide device, this gives a temperature coefficient of approximately 0·4 Å/K.

These temperature effects occur when the operating temperature of a device is changed or when the temperature of the junction region rises during a drive pulse, and for this reason the spectral details are often difficult to resolve.

When devices are operated under carefully controlled temperature and drive conditions, narrow spectral line widths can be observed. These line widths are usually quoted in terms of a frequency variation, and it should be noted that a wavelength of 8500 Å corresponds to a frequency of $3·5 \times 10^{14}$ Hz, and that a line width of 1 GHz corresponds to 0·03 Å. Using heterodyne measuring techniques, line widths of about 0·1 MHz have been reported,[87] but the power available in these narrow line widths is small.

5.8 Modulation effects

The output from a semiconductor laser can be modulated by modulating the current flowing through the junction. The ease with which modulation is thus achieved is an important consideration when the applications of the devices are considered. Thus the response of a laser to rapid variations in the drive current, as well as self-modulation effects in the output, have been extensively studied. A review of this topic is given in Reference 88.

5.8.1 Pulse modulation

When a pulse of current is applied to a laser, a finite delay occurs before laser action is observed. This delay is due to the time required to inject enough electrons to establish the inverted-population conditions necessary for oscillation. The delay decreases as the amplitude of the current pulses I is increased and is given by[89]

$$t = \tau \frac{I}{I - I_0} \tag{5.17}$$

where τ is the spontaneous lifetime of the injected electrons. The value of τ is typically 2 ns, and the delays observed, for currents several times greater than the threshold current, are less than 1 ns.

Considerably longer delays than these are sometimes observed, and values of about 100 ns have been reported with devices operating near room temperature.[90,91] These delays only occur in a minority of devices, and can be minimized by a suitable, but empirical, design of the junction region of the device.

The delays have been explained by a model that postulates that the junction region contains a number of acceptor centres which have several possible energy states.[92,93] In one of these states, the centres absorb radiation and thus inhibit laser action. The absorbing states of the centre are bleached when they trap electrons from the injected current. Thus laser action can occur after a finite delay. It should be noted that, according to this model, laser action is inhibited by optical absorption, even after the inverted population has been established. This is in contrast to an alternative model, in which the establishment of the inverted population is delayed by centres that trap an appreciable number of the injected electrons.

5.8.2   *Microwave modulation*

Once a laser is at or above threshold, it can respond very rapidly to a change in the drive current, and several workers have studied the response of a laser to a current modulated at microwave frequencies.

A laser operating in a continuous mode has been modulated at 11 GHz,[94] but attempts to modulate devices at higher frequencies have some interesting resonance effects.[95]

Experiments have shown that a laser that is oscillating in one of its optical resonant modes of frequency v can be efficiently modulated at a microwave frequency f if the side-band frequencies $v \pm f$ are also resonant optical modes of the cavity. Since the optical modes of the cavity are given by

$$n\lambda = 2\mu L$$

or
$$v = \frac{nc}{2\mu L} \qquad\qquad (5.18)$$

the frequency separation of adjacent modes, and one of the frequencies at which the device can be efficiently modulated, are given by

$$\Delta v = f = \frac{c}{2\mu L} = \frac{c\Delta\lambda}{\lambda^2}. \qquad\qquad (5.19)$$

In the experiments referred to, the laser modes had a wavelength separation of 1·1 Å, and the device was modulated at 46 GHz, in agreement with equation (5.19). These results imply that, to achieve efficient high-frequency modulation, the cavity length must be selected to give the required modulation frequencies.

5.9   Summary

It is convenient to conclude the discussion of semiconductor lasers by summarizing the important properties of some typical devices. As in most of this chapter, the parameters quoted will be for gallium arsenide devices, since these are the only semiconductor lasers that are readily available. These devices are commonly operated with some cooling if the best possible performance is required, or at room temperature if a simple and more convenient system will have an adequate performance.

At low temperatures, diffused and simple epitaxial junction devices give a similar performance. Peak powers up to 50 W and operating duty cycles of 5 per cent. can be obtained with devices cooled to 100 K or below.

At room temperature, heterojunction devices have characteristics superior to those of diffused devices, and have largely superseded the earlier devices. The single-heterojunction structure is best suited for situations where a high peak power is required. With these devices, a peak power of 10 W can be achieved at a duty cycle of 0·5 per cent. For the highest possible duty cycle, the low threshold current of the double-heterojunction structure is an advantage, and with these devices duty cycles of up to 10 per cent. can be readily achieved. However, the beam divergence from the devices is quite large, and the power density at which self damage occurs is lower than in the other devices. Thus the double heterojunction can be used where a high duty cycle is important and a modest peak power is sufficient. Such situations occur, for example, in communications systems.

Table 5.4 summarizes the performance of some lasers that are representative of the range of available devices. Considerably better performances than those quoted have been obtained experimentally. For example, c.w. operation at room temperature has been reported, but the figures given here represent typical devices. In general, the results given can be extrapolated to devices of a different junction area with a proportional change in the drive-current requirements. However, it should be noted that this will change the total power output from the device, but will not change the radiance of the emitting area.

TABLE 5.4. The performance of typical gallium arsenide lasers.

Device structure	Diffused or epitaxial	Single heterojunction	Double heterojunction
Operating temperature (K)	77	300	300
Threshold current density (A/cm^2)	10^3	10^4	2×10^3
Emission wavelength (Å)	8500	9000	9000
Cavity length (cm)	0·1	0·05	0·05
Cavity width (cm)	0·05	0·02	0·01
Threshold current (A)	5	10	1
Operating current (A)	25	25	3
Pulse length (μs)	1	0·2	0·2
Repetition rate (kHz)	50	25	250
Duty cycle (%)	5	0·5	5
Peak power (W)	10	5	1
Mean power (mW)	500	25	50

144

Similarly, it is possible to vary the pulse lengths and repetition rates over quite wide ranges, provided that the operating duty cycle and mean power are kept approximately constant. For example, room-temperature devices may be operated at pulse rates greater than 1 MHz if the pulse length is kept below 50 ns.

Some of the applications of these devices are considered in the next chapter.

5.10 References

1. N. Holonyak and S. V. Bevacqua, *Appl. Phys. Lett.*, **1**, 82 (1962).
2. C. J. Neuse, G. E. Stillman, M. D. Sirikis and N. Holonyak, *Solid State Electron.*, **9**, 735 (1966).
3. J. I. Pankove, H. Nelson, J. J. Tietjen, I. J. Hegyi and H. P. Maruska, *RCA Rev.*, **28**, 560 (1967).
4. J. J. Tietjen, J. I. Pankove, I. J. Hegyi and H. Nelson, *Trans. Met. Soc. AIME*, **239**, 385 (1967).
5. H. Rupprecht, J. M. Woodall, G. D. Petit, J. W. Crowe and H. F. Quinn, *J. Quantum. Electron.*, **4**, 35 (1968).
6. H. Kressell, H. F. Lockwood and H. Nelson, *J. Quantum Electron.*, **6**, 278 (1970).
7. R. N. Hall, G. E. Fenner, J. D. Kingsley, T. J. Soltys and R. O. Carlson, *Phys. Rev. Lett.*, **9**, 366 (1962).
8. M. I. Nathan, W. P. Dumke, G. Burns, F. H. Dill and G. J. Lasher, *Appl. Phys. Lett.*, **1**, 62 (1962).
9. T. M. Quist, R. H. Rediker, R. J. Keyes, W. E. Krag, B. Lax, A. L. McWhorter and H. J. Zeiger, *Appl. Phys. Lett.*, **1**, 91 (1962).
10. K. Weiser and R. S. Levitt, *Appl. Phys. Lett.*, **2**, 178 (1963).
11. G. M. Blom and J. M. Woodall, *Appl. Phys. Lett.*, **17**, 373 (1970).
12. C. Chipaux, G. Durraffourg, J. Loudette, J. P. Noblanc and M. Bernard, in *Radiative Recombination in Semiconductors, Proc. Symp. Paris, 1964*, Dunod, Paris, 1965, p. 217.
13. F. B. Alexander, V. R. Bird, D. R. Carpenter, G. W. Manley, P. S. McDermott, J. R. Peloke, H. F. Quinn, R. J. Riley and L. R. Yetter, *Appl. Phys. Lett.*, **4**, 13 (1964).
14. I. Melngailis, A. J. Strauss and R. H. Rediker, *Proc. IEEE.*, **51**, 1154 (1963).
15. N. Holonyak, D. R. Scifres, R. D. Dupuis and G. W. Zack, *Appl. Phys. Lett.*, **19**, 271 (1971).
16. I. Melngailis, *Appl. Phys. Lett.*, **2**, 176 (1963).
17. I. Melngailis and R. H. Rediker, *J. Appl. Phys.*, **37**, 899 (1966).
18. N. G. Basov, A. V. Dodenkova, A. I. Krasil'nikov, V. V. Nitikin and K. P. Fedoseev, *Fiz. Tverd. Tela.*, **8**, 1060 (1960) [*Sov. Phys. Solid State*, **4**, 847 (1966)].
19. R. J. Phelan, A. R. Calawa, R. H. Rediker, R. J. Keyes and B. Lax, *Appl. Phys. Lett.*, **3**, 143 (1963).
20. J. F. Butler and A. R. Calawa, *J. Electrochem. Soc.*, **54**, 1056 (1956).
21. J. F. Butler, A. R. Calawa, R. J. Phelan, T. C. Harman and A. J. Strauss, *Appl. Phys. Lett.*, **5**, 75 (1964).
22. J. F. Butler, A. R. Calawa, R. J. Phelan, A. J. Strauss and R. H. Rediker, *Solid State Commun.*, **2**, 301 (1964).
23. J. F. Butler and T. C. Harman, *Appl. Phys. Lett.*, **12**, 347 (1968).

24. J. F. Butler and T. C. Harman, *J. Quantum Electron.*, **5**, 50 (1969).
25. J. F. Butler, A. R. Calawa and T. C. Harman, *Phys. Lett.*, **9**, 427 (1966).
26. C. E. Hurwitz, *Appl. Phys. Lett.*, **9**, 116 (1966).
27. F. H. Nicol, *Appl. Phys. Lett.*, **9**, 13 (1966).
28. W. D. Johnston, *J. Appl. Phys.*, **42**, 2731 (1971).
29. O. V. Bogdankevich, M. M. Zvera, A. I. Krasilnikov and A. N. Pechenov, *Phys. Stat. Solidi*, **19**, K5 (1967).
30. C. E. Hurwitz, *J. Quantum Electron.*, **3**, 333 (1967).
31. A. N. Vlasov, G. S. Kosina and O. B. Fedorova, *Zh. Eks. Teor. Fiz.*, **52**, 434 (1967) [*Sov. Phys. JETP.*, **25**, 283 (1967)].
32. N. G. Basov, O. V. Bogdankevich and A. G. Devyathov, *Dokl. Akad. Nauk. SSSR.*, **155**, 783 (1964) [*Sov. Phys. Dokl.*, **9**, 288 (1964)].
33. C. Benoit à la Guillaume and J. M. Debever, *Compt. Rend.*, **261**, 5428 (1965).
34. V. K. Koniukhov, L. A. Kulevskii and A. M. Prokhorov, *Dokl. Akad. Nauk. SSSR.*, **164**, 1012 (1965) [*Sov. Phys. Dokl.*, **10**, 943 (1965)].
35. C. E. Hurwitz, *Appl. Phys. Lett.*, **8**, 243 (1966).
36. C. E. Hurwitz, *Appl. Phys. Lett.*, **8**, 121 (1966).,
37. G. E. Stillman, J. A. Rossi, M. R. Johnson and N. Holonyak, *Appl. Phys. Lett.*, **9**, 268 (1966).
38. A. Z. Grasyok, V. F. Efimkov, I. G. Zubarev, V. A. Katulin and A. N. Mentsev, *Fiz. Tverd. Tela.*, **8**, 1953 (1966) [*Sov. Phys. Solid State*, **8**, 1548 (1966)].
39. V. S. Vavilov and E. L. Nolle, *Dokl. Akad. Nauk. SSSR.*, **164**, 73 (1965) [*Sov. Phys. Dokl.*, **10**, 827 (1966)].
40. N. G. Basov, A. Z. Grasiuk, V. F. Efimkov, I. G. Zubarev, V. A. Katulin and Yu M. Popov, in *The Physics of Semiconductors*, *Proc. Intl. Conf. Kyoto, 1966*, The Physical Society of Japan, Tokyo, 1966, p. 277.
41. R. Dingle, R. B. Zetterstrom, K. L. Shaklee and R. F. Leheny, *Appl. Phys. Lett.*, **19**, 5 (1971).
42. N. G. Basov, O. V. Bogdankevich, A. N. Pechenov, G. B. Abdulaev, G. A. Akhundov, and E. Yu Salaev, *Dokl. Akad. Nauk. SSSR.*, **161**, 1059 (1965) [*Sov. Phys. Dokl.*, **10**, 329 (1965)].
43. N. G. Basov, O. V. Bogdankevich, P. G. Eliseev and B. Lavruskin, *Fiz. Tverd. Tela.*, **8**, 1341 (1966) [*Sov. Phys. Solid State*, **8**, 1073 (1966)].
44. C. E. Hurwitz and R. J. Keyes, *Appl. Phys. Lett.*, **5**, 139 (1964).
45. N. G. Basov, A. Z. Grasyuk and V. A. Kautlin, *Dokl. Akad. Nauk. SSSR.*, **161**, 1306 (1965) [*Sov. Phys. Dokl.*, **10**, 343 (1965)].
46. C. Benoit à la Guillaume and J. M. Debever, *Compt. Rend.*, **259**, 2200 (1964).
47. C. Benoit à la Guillaume and J. M. Debever, *Solid State Commun.*, **2**, 145 (1964).
48. I. Melngailis, *J. Quantum Electron.*, **1**, 104 (1965).
49. C. Benoit à la Guillaume and J. M. Debever, in *Radiative Recombination in Semiconductors*, *Proc. Symp. Paris, 1964*, Dunod, Paris, 1965, p. 255.
50. R. J. Phelan and R. H. Rediker, *Appl. Phys. Lett.*, **6**, 70 (1965).
51. G. K. Avenkieva, N. A. Goryunova, V. D. Prochukhan, S. M. Ryvkin, M. Serginov and Yu G. Shreter, *Fiz. Tekh. Poluprov. SSSR.*, **5**, 174 (1971) [*Sov. Phys. Semicond.*, **5**, 151 (1971)].
52. F. M. Berkovskii, N. A. Goryunova, V. M. Orlow, S. M. Ryvkin, V. I. Sokolova, E. V. Tsvetkova and G. P. Shpen'kov, *Fiz. Tekh. Poluprov. SSSR.*, **2**, 1218 (1968) [*Sov. Phys. Semicond.*, **2**, 1027 (1969)].
53. S. G. Bishop, W. J. Moore and E. M. Swiggard, *Appl. Phys. Lett.*, **16**, 459 (1970).
54. C. Benoit à la Guillaume and J. M. Debever, *Solid State Commun.*, **3**, 19 (1965).
55. I. Melngailis and A. J. Strauss, *Appl. Phys. Lett.*, **8**, 179 (1966).
56. C. H. Hurwitz, *J. Quantum Electron.*, **1**, 102 (1965).

57. J. O. Dimmock, I. Melngailis and A. J. Strauss, *Phys. Rev. Lett.*, **16**, 1193 (1966).
58. R. O. Carlson, *J. Appl. Phys.*, **38**, 661 (1966).
59. H. Nelson, *RCA Rev.*, **24**, 603 (1963).
60. M. H. Pilkuhn and H. Rupprecht, *J. Appl. Phys.*, **38**, 5 (1967).
61. I. Hayashi, M. Panish and P. W. Foy, *J. Quantum Electron.*, **5**, 211 (1969).
62. H. Kressel and H. Nelson, *RCA Rev.*, **30**, 106 (1969).
63. I. Hayashi and M. B. Panish, *J. Appl. Phys.*, **41**, 150 (1970).
64. H. Kressel, H. Nelson and F. Z. Hawrylo, *J. Appl. Phys.*, **41**, 2019 (1970).
65. H. Kressel, H. F. Lockwood and H. Nelson, *J. Quantum Electron.*, **6**, 278 (1970).
66. Zh I. Alferov, V. M. Andreev, E. L. Portnoi and M. K. Trukan, *Fiz. Tekh. Poluprov SSSR.*, **3**, 1328 (1969) [*Sov. Phys. Semicond.*, **3**, 1107 (1970)].
67. M. B. Panish, I. Hayashi and S. Sumski, *Appl. Phys. Lett.*, **16**, 326 (1970).
68. H. Kressel and F. Z. Hawrylo, *Appl. Phys. Lett.*, **17**, 169 (1970).
69. I. Hayashi, M. B. Panish and F. K. Reinhart, *J. Appl. Phys.*, **42**, 1929 (1971).
70. Zh I. Alferov, V. M. Andreev, D. Z. Garbuzov, Yu V. Zhilyaev, E. P. Morozov, E. L. Portnoi and V. G. Trofin, *Fiz. Tekh. Poluprov. SSSR.*, **4**, 1826 (1970) [*Sov. Phys. Semicond.*, **4**, 1573 (1971)].
71. B. I. Miller, J. E. Ripper, J. C. Dyment, E. Pinkas and M. B. Panish, *Appl. Phys. Lett.*, **18**, 403 (1971).
72. H. Kressel, H. F. Lockwood, and F. Z. Hawrylo, *J. Appl. Phys.*, **43**, 561 (1972).
73. G. H. B. Thompson, GaAs Laser Symposium, Cardiff (1972).
74. W. Engler and M. Garfinkel, *J. Appl. Phys.*, **35**, 1734 (1964).
75. J. C. Dyment, *Appl. Phys. Lett.*, **10**, 84 (1967).
76. R. F. Broom, C. H. Gooch, C. Hilsum and D. J. Oliver, *Nature*, **198**, 4878 (1963).
77. M. Pilkuhn and H. Rupprecht, *Proc. IEEE.*, **51**, 1243 (1963).
78. M. Pilkuhn and H. Rupprecht, *J. Appl. Phys.*, **38**, 5 (1967).
79. P. Selway and A. R. Goodwin, *J. Phys. D: Appl. Phys.*, **5**, 904 (1972).
80. M. J. Adams and M. Cross, *Solid State Electron.*, **14**, 865 (1971).
81. J. R. Biard, W. N. Carr and B. S. Reed, *Trans. AIME.*, **230**, 286 (1964).
82. M. Clifton and P. R. Debye, *Appl. Phys. Lett.*, **6**, 120 (1965).
83. J. C. Dyment and L. A. D'Asaro, *Appl. Phys. Lett.*, **11**, 292 (1967).
84. I. Hayashi, M. B. Panish, P. W. Foy and S. Sumski, *Appl. Phys. Lett.*, **17**, 109 (1970).
85. D. P. Cooper, C. H. Gooch and R. J. Sherwell, *J. Quantum Electron.*, **2**, 329 (1966).
86. M. Ettenberg, H. J. Sommers, H. Kressel and H. F. Lockwood, *Appl. Phys. Lett.*, **18**, 571 (1971).
87. J. W. Crowe and R. M. Craig, *Appl. Phys. Lett.*, **5**, 72 (1964).
88. T. L. Paoli and J. E. Ripper, *Proc. IEEE.*, **58**, 1457 (1970).
89. K. Konnerth and C. Lanza, *Appl. Phys. Lett.*, **4**, 120 (1964).
90. N. N. Winogradoff and H. K. Kessler, *Solid State Commun.*, **2**, 119 (1964).
91. K. Konnerth, *IEEE. Trans. Electron Dev.*, **12**, 506 (1965).
92. G. Fenner, *Solid State Electron.*, **10**, 753 (1967).
93. J. E. Ripper and J. C. Dyment, *J. Quantum Electron.*, **5**, 391 (1969).
94. B. S. Goldstein and R. M. Weigand, *Proc. IEEE.*, **53**, 195 (1965).
95. S. Takamiya, F. Kitosawa and J. I. Nishizawa, *Proc. IEEE.*, **56**, 135 (1968).

6

The Applications of Electroluminescent Diodes and Lasers

6.1 Introduction

Semiconductor electroluminescent devices possess a number of features, including small size, high reliability and high luminance, that give them considerable advantages when compared with other sources of radiation, such as tungsten-filament lamps. They also possess other features, such as the ability to be modulated at high frequencies, that enable them to satisfy requirements that cannot be met by other devices.

In considering the applications of electroluminescent devices, it will be convenient to divide these into two categories, depending on the spectral region in which the devices emit.

[i] *Sources emitting visible radiation.* This category includes the use of devices as indicator lamps and in display systems. In these applications, a high modulation speed is not required, but the important features are high reliability and luminance. In many situations it is also desirable that devices should operate at power levels that are compatible with transistor circuits.

[ii] *Infrared sources.* This category includes the use of incoherent sources and lasers, since the applications of these devices are often qualitatively similar. The main difference to be considered is the higher radiance of lasers, which can give a better system performance. This advantage must be balanced against the increased complexity of the system that is usually required to operate a laser.

6.2 General considerations

6.2.1 Reliability

Semiconductor devices are rugged and can be expected to have a high reliability. Electroluminescent diodes are no exception, and devices operating under normal conditions have a 'life' of 10^5 h or more.[1,2] The 'life' which is quoted is the time during which the efficiency of a device falls by a factor of two, and it is important to note that the devices do not normally show a sudden catastrophic failure, but degrade at a steady and predictable rate. In this respect they have a considerable advantage over filament lamps.

The life of lasers still leaves room for considerable improvement. Devices operating under pulsed conditions, where the duty cycle is about 1 per cent., show lives of 10^3 h or more, and this is adequate for many applications. However, devices operating continuously at room temperature have not yet shown a life significantly greater than 100h.

The degradation of electroluminescent devices is receiving considerable attention, and several possible mechanisms have been proposed. The two which explain many of the observed features are as follows.

According to a mechanism proposed by Longini,[3] interstitial impurity atoms exist in appreciable concentrations in the material adjacent to the p–n junction, but are prevented from entering the junction region by the junction field. Application of a forward bias to the diode lowers the junction field, so that the interstitial atoms can diffuse into the junction region where they act as centres for non-radiative or space-charge recombination.

The mechanism proposed by Gold and Weisberg[4] is known as the 'phonon-kick' mechanism. In this, a non-radiative recombination process imparts enough energy to an atom to remove it from its lattice site to an interstitial position. This then gives rise to non-radiative recombination centres, which degrade the device efficiency.

Both mechanisms predict that the current through a device under constant bias will increase as degradation proceeds, in agreement with experimental observations. However, the Longini mechanism, which depends on a diffusion rate, should show a strong temperature dependence, whereas the phonon-kick mechanism shows a temperature dependence that depends on the ratio of the non-radiative and radiative current components. At present, the experimental evidence is not sufficient to establish the importance of these mechanisms in practical devices.

Another form of failure can be observed in semiconductor lasers.[5,6] This occurs when the optical flux density through the face of the device exceeds a critical value and results in the formation of pits or other visible damage to the face. The output-power levels at which this occurs depend somewhat on the structure of the device and the operating conditions, but in general damage may be observed at output-power levels exceeding 500 W per centimetre of junction width. Since the narrow dimension of the laser source is about 2 μm, this corresponds to a power density greater than 10^6 W/cm^2. This form of damage, which results in a catastrophic decrease in the output of a device, can be avoided by restricting the output to values below this level.

6.2.2 Radiance and luminance

In many situations the high luminance or radiance of a semiconductor device is an important consideration. For example, a diode with an efficiency of 1 per cent. operating at a current density of 5 A/cm^2 with an applied voltage of 2 V will have a radiance of approximately 10 mW/cm² sr. If the device were pulsed to a current density of 5000 A/cm², the radiance would be 10 W/cm² sr.

Much higher values of radiance are achieved with lasers because of the small source size and the relatively small angle of emission. For example, a laser operating at a current density of 5000 A/cm² with a junction area of 10^{-2} cm² would emit 10 W from a source area of 0·1 cm × 2 μm into a solid angle of about 0·25 sr. This device would thus have a pulsed radiance of about 2×10^6 W/cm² sr.

The luminance of devices emitting visible radiation is also high. Taking the previous example of a diode with an efficiency of 1 per cent. and a luminous efficiency of radiation of 15 lm/W, the device luminance would be 4500 cd/m² or 1500 ft L. In practice, values as high as this can be readily achieved, and the luminance of such devices then compares favourably with that of a diffusing surface exposed to strong sunlight.

6.2.3 Modulation

Electroluminescent diodes can be modulated at high frequencies, and this makes them valuable in communications and radar systems. The frequency limitation depends on the nature of the device. For example, devices based on the direct band-gap materials GaAs and $Ga_xAs_{1-x}P$ show rise times of about 5 ns, and can be efficiently modulated at frequencies greater than 100 MHz. At the other extreme, red-emitting GaP devices have a rise time of about 200 ns and a correspondingly lower maximum modulation frequency.

This high speed of response is seldom utilized in 'visible' electroluminescent devices, where the human eye is the relevant detector. However, it is useful to note that a semiconductor lamp will exhibit its maximum efficiency even when subjected to a short drive pulse, whereas a filament lamp, whose efficiency increases rapidly with temperature, would be inefficient under similar conditions.

The maximum modulation speed of lasers is also high, but there is a finite delay between the application of a current pulse and the onset of stimulated emission. This is the time required to establish an inverted population, and varies inversely with the amplitude of the current pulse. If the highest modulation frequencies are required, it may be necessary to bias the device with a current close to the threshold value and superimpose the modulating current on this.

By using a laser in this way the time necessary to establish the inverted population can be almost entirely obviated, and the modulation frequency of the laser is then limited by the stimulated lifetime of injected carriers. Thus pulse lengths of 10–100 ps and modulation frequencies of 10–100 GHz are possible. However, as yet no practical systems have exploited this potential.

6.3   Film annotation

One of the earliest viable applications of visible electroluminescent diodes was their use in film annotation. This application arises when a series of photographs is taken, and it is necessary to record simultaneously on the

FIGURE 6.1. Film annotation using electroluminescent diodes

film information relating to those photographs (Figure 6.1). This is particularly important, for example, in aerial photography for survey purposes.

A small array of lamps is built into the camera so that the array is held close to the film. The required information is recorded on the film in binary code by exposing it with a selection of the lamps.

The first devices to be used for this purpose were red-emitting gallium phosphide diodes, even though, at the time, these diodes had efficiencies which were only 0·01 per cent. Although the efficiency of the lamps was low, an adequate film exposure could be obtained by pulsing the lamps with a 100 mA, 1 ms current pulse. With the improved efficiency of lamps and improvements in the techniques used to make small arrays, all the requirements for this type of system can be readily met.

Film annotation is a very simple application of electroluminescent devices as it is only necessary to discriminate between the presence and absence of an exposure to record binary-coded information. Analogue information can also be recorded. This can be achieved by using an analogue signal to modulate the radiation from an electroluminescent diode. This radiation is then focused onto a moving film to give a trace of variable density. In even more sophisticated systems the focused radiation can be made to traverse the film in two directions and thus give a two-dimensional record.

6.4 Indicator lamps

One of the most obvious uses of electroluminescent diodes is to indicate the functioning of equipment and circuits. Here semiconductor devices are in direct competition with tungsten filament lamps.

The luminous efficiency of small tungsten filament lamps is about 10 lm/W, but this is reduced by at least an order of magnitude when the radiation is filtered to give a coloured lamp. The luminous efficiency of electroluminescent devices is about 0·3 lm/W, and this compares favourably with a tungsten lamp plus a filter.

The high reliability of electroluminescent devices is another major consideration, and the higher initial cost of such lamps is often more than compensated by the saving which results. Here it must be remembered that the cost of replacing a lamp must include not only the cost of the lamp, but also the cost of labour involved in making the replacement. Moreover, since electroluminescent diodes degrade rather than fail suddenly, they can be replaced at a convenient time which does not disrupt the functioning of the system in which they are installed.

Electroluminescent diodes can also find wide application as state indicators in logic circuits. Here it is often desirable to have a visual indication of the state of a logic circuit. A circuit can often provide enough power to operate an electroluminescent diode, since a drive current as low as 1 mA will often give sufficient visual output. In other situations, an electroluminescent diode can be integrated with a simple drive circuit designed to amplify the power available from the logic circuit. For example, the device shown in Figure 6.2

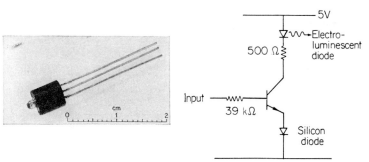

FIGURE 6.2. A logic-indicating device (Crown Copyright, reproduced by permission of the Controller, Her Majesty's Stationery Office)

contains a diode and transistor drive circuit. This device is designed to be wired into logic circuits and impose a negligible load. In the 'on' state (2·5 V), a current of less than 30 μA is drawn by the input, and this is amplified so that the electroluminescent diode operates at 4 mA and gives an output that is clearly visible in practical situations.

F

6.5   Display systems

Possibly the largest likely area of application of electroluminescent devices is their use in display systems in which it is required to make a visual display of information. Such a display would find use, for example, in a computer terminal or in digital instrumentation.

In these applications electroluminescent devices are in competition with a number of alternative approaches. When large quantities of data are to be displayed a cathode-ray tube is often the most viable display. With a c.r.t. display about 10^5–10^6 information points are available and up to 100 rows of characters can be displayed. To match this 10^5 or more light-emitting diode elements would have to be used, and this is unlikely to be an economic proposition.

Many applications, however, require the display of a small number of characters. For example, a digital voltmeter might use five numerals. These are often gas-discharge devices in which one device contains a stack of characters, one of which is energized. The stack of characters occupies an appreciable thickness and the number of characters that can be displayed is limited. Electroluminescent devices are ideally suited to this type of application.

The display devices, illustrated in Chapter 4, consist of a number of light-emitting elements. A character is displayed on such a device by driving a suitable selection of these elements.

The first stage in the design of a display system is to determine the range of characters that are needed and the type of array necessary to give an adequate representation of these characters.

The second problem is to design addressing circuits that will accept binary-coded information and drive the appropriate lamps. These addressing circuits can conveniently be considered in three parts (Figure 6.3). The first

FIGURE 6.3. Block diagram of a display system

of these involves decoding the binary-coded information to identify the symbol which is to be displayed. The second stage, which is known as encoding or character generation, is the selection of the lamps that are required to display the symbol. The third stage is the drive circuits required to pass current through the selected lamps. The design of these circuits will be considered in Sections 6.4.2–6.4.4.

The design of addressing circuits relies, to a large extent, on the utilization of the wide range of digital logic circuits that the technology of silicon integrated circuits has made available. To consider this problem, it is

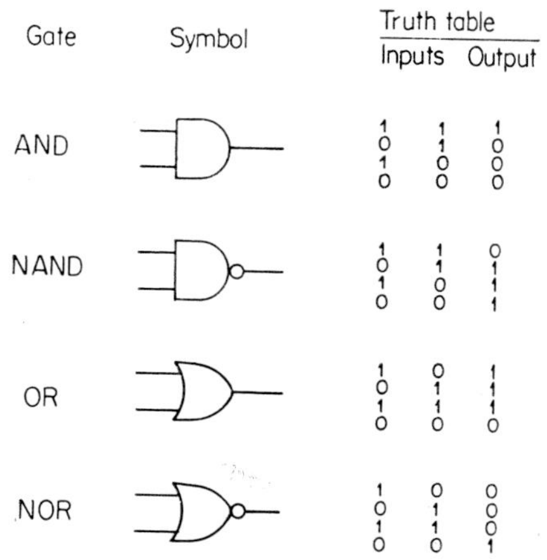

Gate	Symbol	Inputs		Output
AND		1 0 1 0	1 1 0 0	1 0 0 0
NAND		1 0 1 0	1 1 0 0	0 1 1 1
OR		1 0 1 0	0 1 1 0	1 1 1 0
NOR		1 0 1 0	0 1 1 0	0 0 0 1

FIGURE 6.4. Basic logic gates

necessary to make use of the terminology of logic circuits and understand the functioning of some of the basic circuits that are used. These are illustrated in Figure 6.4. Positive logic, in which the high level represents the '1' or 'true' state and the low level represents '0' or 'false' is used here. These gate circuits have two (or more) parallel inputs and one output. The output of a gate will normally be '0' unless certain input conditions are fulfilled, in which case the output will be '1'. If the output is qualified by the symbol O, the output is inverted and the significant state is then '0'.

The gates shown in Figure 6.4 function as follows:

(a) AND gate: the output is '1' if *all* the input levels are '1'; otherwise the output is '0'.
(b) NAND gate: the output is '0' if *all* the input levels are '1'; otherwise the output is '1'.
(c) OR gate: the output is '1' if *any one* of the input levels is '1'; otherwise the output is '0'.
(d) NOR gate: the output is '0' if *any one* of the input levels is '1'; otherwise the output is '1'.

A NAND gate with a single input can be used to represent the inversion of a logic state (1 → 0 or 0 → 1) where this is required.

6.5.1 Display format

One of the simplest display formats, but one which is suitable for the display of numerals only, consists of seven bars arranged as in Figure 6.5(a). With this format it is possible to give an unambiguous display of the numerals 0–9 and some letters, and the style of the display is improved if the fount is italicized as shown in Figure 6.5(b).

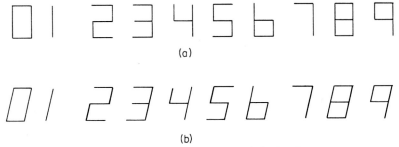

(a)

(b)

FIGURE 6.5. Seven-segment numeral display format

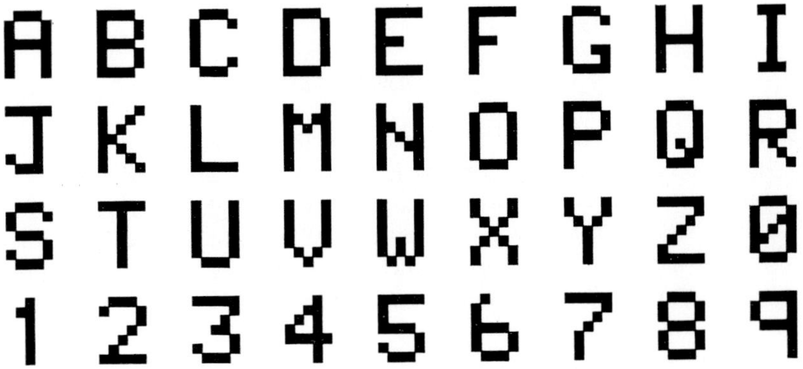

ABCDEFGHI
JKLMNOPQR
STUVWXYZ0
123456789

FIGURE 6.6. 7×5 matrix display format

A more satisfactory format is shown in Figure 6.6. This consists of an array of 35 lamps arranged in a 7×5 matrix. For many purposes this gives an adequate display of the numbers and letters, and can also display many of the commonly used symbols. An even better display can be obtained with a 9×7 matrix, but this is rarely needed.

6.5.2 Decoding

Information which is to be displayed will normally be presented in binary-coded form. Using a 6-bit code, 64 ($= 2^6$) symbols can be identified, and this can cover not only the alphabet and numerals, but also a range of conventional symbols. A typical decoding circuit will have a six-line input and a 64-line output. Any combination of parallel inputs will activate one specific output line to identify a character.

A decoding circuit can be constructed from a number of AND gates, as shown in Figure 6.7 for the simple case of a 3-bit input and an eight-line output.

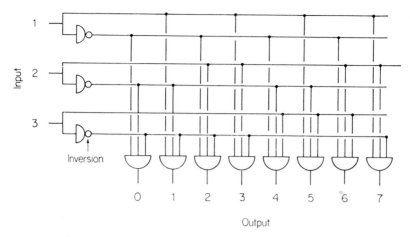

FIGURE 6.7. A 3-bit decoding circuit using AND gates

Integrated circuits that will decode a 4-bit input to a sixteen-line output are now readily available. Four of these can be combined to decode a 6-bit input as shown in Figure 6.8. The first four significant digits are fed in parallel to each of the 4-bit decoders. The remaining two input digits are decoded to four lines that are then used to select one of the 4-bit decoders by driving the 'enable' input of these decoders.

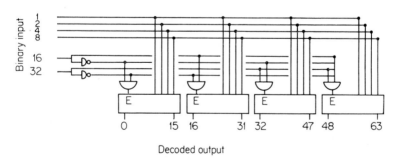

FIGURE 6.8. A 6-bit decoder using four 4-bit decoders

6.5.3 Encoding or character generation

The purpose of encoding circuits is to determine which lamps are to be on to display a particular character. These circuits thus determine the style or fount of the symbols, and are thus often called 'character generators'.

A typical encoding circuit will have a number of parallel inputs, one of which will be activated to identify a character, and a number of output lines, several of which will be activated in parallel to energize the required diodes.

The simplest form of encoding circuit consists of a diode matrix, as shown in Figure 6.9(a). To display S symbols using L lamps, a matrix of $S \times L$ crossover points is required. On average, each symbol requires about 40 per cent. of the lamps to be on, so that the number of diodes in an encoder matrix is approximately $0 \cdot 4 \, SL$.

Some economy in this number can be effected by grouping lamps so that

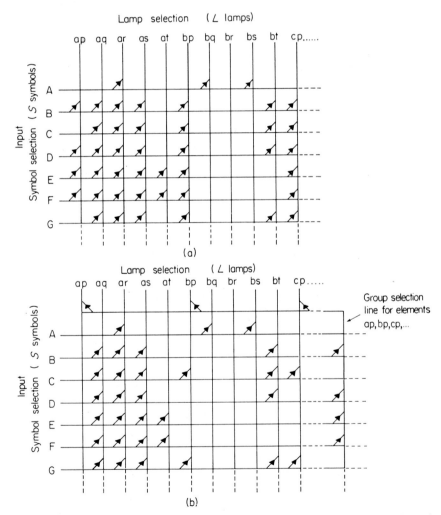

FIGURE 6.9. A diode encoding matrix with parallel output: (a) basic circuit and (b) circuit using group selection for the characters B, D, E, F, . . .

those that are frequently on at the same time can be driven together. For example, in a 7×5 format, the eleven characters B, D, E, F, H, K, L, M, N, P and R all require the same vertical column of seven lamps. In a simple encoding matrix this would require $7 \times 11 = 77$ diodes. However, by driving a group of lamps as shown in Figure 6.9(b), this function can be achieved using one diode for each character (11) and each lamp (7). Thus only $7+11 = 18$ diodes are required for this part of the encoding matrix.

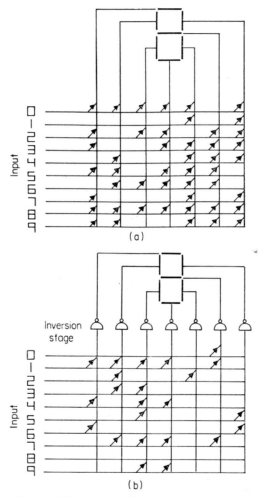

FIGURE 6.10. An encoding circuit for a seven-bar numeral: (a) an input signal switches elements on and requires 47 diodes and (b) an input signal switches elements off and requires 23 diodes

In a similar way, it is possible to identify some lamps that are on more frequently than they are off. In this situation, an economy in the encoding matrix can be achieved by arranging for these lamps to be on except when not required. This will require an inversion stage in the drive circuit to these particular lamps. For example, in a seven-bar numeral display, the lower right-hand-side vertical element is on for all numbers except 2, and, similarly, each lamp is used for at least six of the ten numerals. The economy that results from switching lamps off instead of on is illustrated in Figure 6.10, where the number of diodes in the encoding matrix is reduced from 47 to 23. In this example, the drive circuits all have to be inverted, but this is not necessarily an additional complication, as transistor drive circuits may accomplish this with no additional components.

The encoding circuits that have been described have one output line for each lamp of the array and, for an array of 35 lamps or more, the number of leads required becomes inconveniently large. An alternative form of encoding circuit overcomes this problem by arranging for the encoding information to be delivered in series or series–parallel form. For example, the circuit shown in Figure 6.11 has five output lines corresponding to the

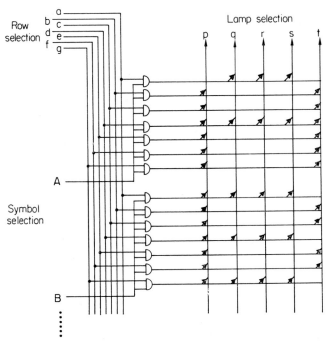

Figure 6.11. An encoding matrix with a series–parallel output

five columns of a 7×5 array. These five output lines give the states of any row that is selected by one of the seven row input lines. A simultaneous input to one of the row lines and one of the character lines will cause the output lines to give the appropriate information. To obtain the full speci- fication of a character, all the row input lines must be selected in turn so that the output lines give, in sequence, the states of row 1, 2, ..., 7. The way in which such an output is used to drive an array will be considered in Section 6.5.4.

The encoding circuits or character generators that have been discussed in the preceding paragraphs are a type of read-only memory (r.o.m.). Integrated circuits that have been designed to perform these functions are now widely obtainable, and these usually combine several of the addressing operations. For example, one integrated circuit package might combine decoding and encoding functions, and could also accept row information in binary-coded form. The total number of leads to such a circuit, excluding power supplies, would be 14 (comprising 6-bit character selection, 3-bit row selection and five-line row output) compared with a possible 99 (64 characters plus 35 elements) for a parallel input–output device.

6.5.4   *Drive stages*

[i] *Array classification.* The final stage in a display system is the application of information from the encoder to the display matrix to operate the lamps required to form the specified character. The design of this stage will depend on the form of the encoder output and the input requirements of the display. In general, the circuits discussed here can be used to drive simple displays, such as the seven-bar numeric, but, to illustrate more fully the principles involved, the more complex case of a 35-element alphanumeric display will be considered. It will be convenient to designate the rows of such an array as a ... g and the columns p ... t.

Two types of array can be identified, according to the way in which the electrical connections are made to the elements. The basic structure of these has been described in Chapter 3.

[a] Parallel-input (common-anode or -cathode) arrays. In this form of array, one side of each element is connected to a common terminal, and separate leads are taken to the other side of each element. For a matrix of $m \times n$ elements, the number of leads required is $mn + 1$. To drive any selection of lamps in the array, it is merely necessary to apply a stabilized current to the appropriate leads.

[b] Coordinate-input arrays. In this format, connections are made to the lamp anodes in rows and to the cathode in columns (or vice versa). Thus, for a matrix of $m \times n$ elements, only $m + n$ leads are required. However, this

simplification imposes constraints on the way in which the display can be operated. For example (Figure 6.12), lamp dq can be driven by applying drive signals of opposite polarity to row d and column q; similarly, lamp er is operated by driving row e and column r, and so on. This form of drive operates only because the array elements have a diode characteristic, so that the current paths parallel to dq through, for example, elements dt, te and eq in series, have a high impedance (across the reverse-biased element te) and thus do not conduct.

If two elements, such as dq and er, are both to be displayed, a parallel drive to rows d and e and to columns q and r would cause unwanted elements dr and eq also to be driven. This can only be overcome by scanning the array to drive the elements either sequentially or one row or column at a time.

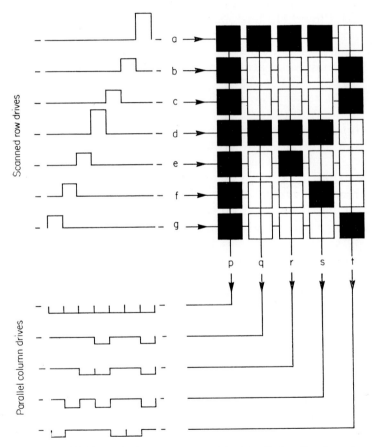

FIGURE 6.12. The scanning sequence required to display the character R on a coordinate matrix (pulse amplitudes represent the current flow)

The most widely used technique is to scan the array row by row. The rows are driven in sequence a, b, . . . , and, at the appropriate time, the correct selection of columns is driven in parallel. For example, the drive sequence required to display the character R is shown in Figure 6.12.

In an array which is scanned in this way, the elements can only be on for a duty cycle corresponding to the number of rows or columns to be scanned. Thus, to maintain the required output, the elements must be operated at a high peak current. For a row-scanned array, the row drive must be capable of driving up to five lamps in parallel at a peak current I_p of seven times the required mean value I_m, and the column drives must each provide a current of $7 I_m$. In order that each element shall take the same current, the column drives must be stabilized at a value of I_p, but the row drives must be unstabilized and supply a current of up to $5 I_p$.

These requirements are summarized in Table 6.1.

TABLE 6.1. The drive requirements for 35-element displays. It is assumed that an element requires a mean current of I_m to give adequate luminance.

Array type	Number of leads	Maximum drive requirements per lead		
		Peak current	Mean current	Stabilization required
Parallel input	35	I_m	I_m	Current
Coordinate input	7 row 5 column	$5 \times 7I_m$ $7I_m$	$5I_m$ $7I_m$	Voltage Current

[ii] *Memory requirements.* The problem of operating lamps at low duty cycles and high peak current could be overcome if the display elements could be switched to the 'on' state by the scanning circuits and then remain on, deriving power from a separate supply. For example, an electroluminescent device with a bistable current–voltage characteristic of the form shown in Figure 6.13(a) would fulfil the required function. This device would be biased to the point V_h, but would only be switched on if pulsed with a voltage greater than V_0. To use such a device, a modified scanning circuit such as shown in Figure 6.13(b) would be necessary. Here voltage pulses $\pm V$ such that

$$\tfrac{1}{2}(V_a - V_h) < V < (V_a - V_h)$$

are applied simultaneously to a row and column to switch an element on. The element is switched off by momentarily reducing the bias voltage to zero.

A bistable characteristic of the form shown in Figure 6.13 can be obtained

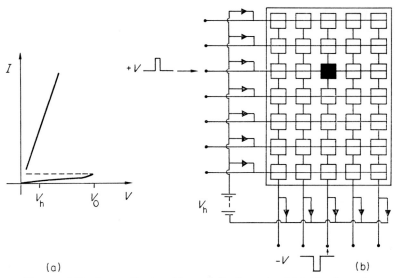

I

$+V$

V_h V_0 V

V_h

$-V$

(a)   (b)

FIGURE 6.13. A coordinate-addressed display using bistable elements

by making the electroluminescent diode one junction of a *p–n–p–n* switch. This approach has been demonstrated experimentally, but at present no practical devices are available, and the required memory properties must be provided by additional circuit elements located with the display element in the matrix. This also is not a practical approach to the problem.

In some situations, a convenient form of memory is provided by a shift register. A typical circuit (Figure 6.14) has signal and clock inputs and a number of parallel outputs. When the clock input receives a pulse, the output states all shift along one stage, and the input signal is read into the first stage. The shift register can thus store a number of bits of information, and the output powers available can be sufficient to operate several lamps in parallel.

[iii] *Output stages.* In many circumstances the output power available from the logic encoding circuits is insufficient to drive the display elements at the required level, and in other circumstances the logic output may be of the wrong polarity. It may thus be necessary to amplify and invert the encoder outputs. In doing this it must be remembered that the display elements are diodes with a low series resistance, and the circuit must supply a stabilized current to the device. These functions are easily achieved by means of simple transistor circuits.

[iv] *Encoder–matrix combinations.* It is now necessary to consider how the two basic types of encoder circuit described in Section 6.5.3 are used in conjunction with the parallel-input and coordinate-input arrays. There are thus four such combinations to be considered.

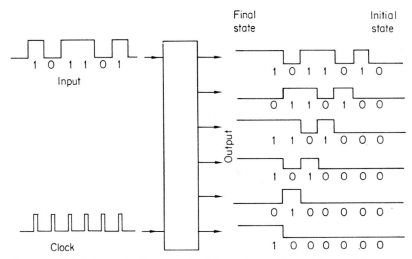

FIGURE 6.14. Schematic diagram of a shift register with a 6-bit parallel output; the successive output states for the input sequence 1 0 1 1 0 1 are shown

[a] Parallel-output encoder, parallel-input array. In concept, the simplest encoder–matrix system is that in which each output from a parallel-output encoder drives one of the lamps of a common-cathode (or -anode) array. In this system, it is merely necessary that the encoder outputs should switch a stabilized current to the lamps that are required to be on to form the desired character (Figure 6.15).

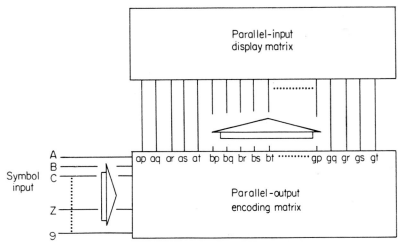

FIGURE 6.15. A display system using a parallel-output encoder with a parallel-input array

[b] Parallel-output encoder, coordinate array. A parallel-output encoder can be used to drive a coordinate array as shown in Figure 6.16. In this circuit, seven of the 35 encoder outputs drive each of the five array columns through 35 AND gates; the second input to these gates is derived from the circuit which scans the array row by row. At each stage of the row-scanning

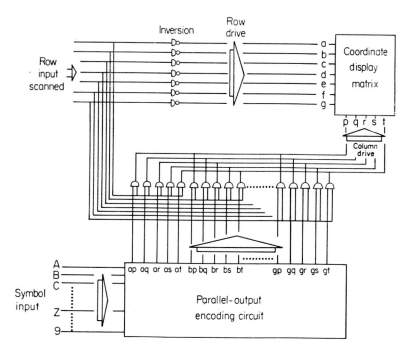

FIGURE 6.16. A display system using a parallel-output encoder with a coordinate-input array

cycle, the appropriate gates are opened to allow the encoder outputs corresponding to that row of lamps to drive the array columns. For example, at row scan position a, gates ap, . . . , at are open and thus encoder outputs ap, . . . , at operate on columns p, . . . , t. At the next scan position, gates bp, . . . , pt are open, and encoder outputs bp, . . . , pt operate the columns. Thus the display is built up by the sequence a, . . . , g at a rate sufficient to avoid the appearance of flicker.

An inversion stage is shown in the row drive of the circuit given in Figure 6.17. This is included because the rows and columns must be driven with opposite polarities. In practice this inversion may be necessary in either the row or column drives, depending on the polarity of the array connections. The required inversion and the necessary row and column current drives, as specified in Table 6.1, are easily achieved using simple transistor switches.

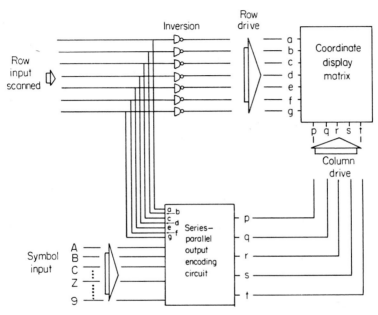

FIGURE 6.17. A display system using a series–parallel encoder with a coordinate-input array

[c] Series–parallel output encoder, coordinate array. The use of a series–parallel output encoder to drive a coordinate matrix is shown in Figure 6.17. Here the column outputs of the encoder are connected to the array columns, and the row input to the encoder is derived from the row-scanning switch.

In practice the encoding or character-generating circuit used here is likely to be an m.o.s. integrated circuit with binary-coded inputs. Thus the row input must be taken from a suitable binary counter that is derived from the row scan. Alternatively, a binary counter can be used to give the encoder row input and must also be decoded to give the required row-drive scan.

As in the previous circuit the correct amplitude and polarity of the currents applied to the array are obtained from simple transistor switching circuits.

[d] Series–parallel output encoder, common-anode array. When a series–parallel output encoder is used to drive an array, the array must either be driven row by row, so that each lamp can only be on for a low duty cycle, or a memory must be provided to allow lamps to stay on while the encoder is being cycled through the seven-row scan sequence. Two ways in which these alternatives can be accomplished are shown in Figures 6.18 and 6.19.

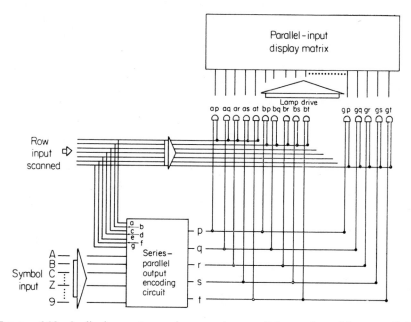

FIGURE 6.18. A display system using a series–parallel encoder with a parallel-input array

In the system shown in Figure 6.18, 35 two-input AND gates are used so as, effectively, to convert the parallel-input array into a coordinate array, and, since the array must now be scanned row by row, the advantages of the parallel-input array are lost.

Each lamp is driven by a two-input gate, so that the gate is opened by the scanning switch only when the encoder is giving information appropriate to that lamp.

In the system shown in Figure 6.19, shift registers are used to store the information from the encoder and drive the lamps in the array. A sequence of seven pulses fed through a ring counter drives the row input of the encoder. The encoder outputs are applied to five 7-bit shift registers, and information is fed into these registers by the application of a train of clock pulses. A delay introduced into the clock circuit ensures that the registers only shift when new information is available from the encoder, and it should be noted that the operation of this circuit requires only one sequence of seven pulses to 'load' the shift-register memories. The encoder circuits can thus be used to drive several arrays in sequence.

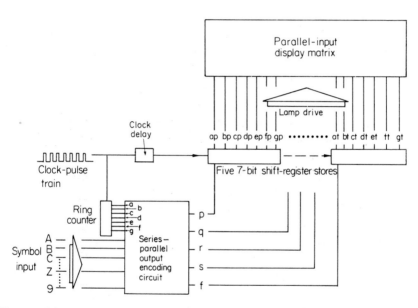

FIGURE 6.19. A display system using a series–parallel encoder with a parallel-input array and shift-register stores

6.5.5 Multicharacter displays

The circuits shown in Figures 6.15–6.19 illustrate some of the ways in which a single array of lamps can be operated in conjunction with an encoding circuit. In practice, the design of these circuits will be influenced by the functions performed by available integrated circuits and, since these are relatively expensive, it may be desirable to use one encoding circuit to address a number of arrays. Although a detailed study of this problem is beyond the scope of this book, some general principles can be outlined, and a more detailed discussion can be found in Reference 7.

In general, the input to a multicharacter display system will be a series of 6-bit words that identify the characters that are to be displayed. Since each word is only available for a short time, the information which they contain must be stored in order to display the characters for as long as required. This storage can be used either after or before the character-generator circuits as shown in Figure 6.20.

In the first approach, each array has a 35-bit memory associated with it, and can thus operate continuously. Such a system can be used with both parallel-input and coordinate-input arrays, and the basic circuits are similar to those shown in Figures 6.15 and 6.16, with the encoding circuits replaced by a 35-bit memory. Each memory is loaded by the output of the encoder which is shared in sequence between several arrays. The limitation on the number of arrays which can be used depends on the speed at which the encoding and memory circuits can operate in relation to the speed at which it is desired to change or update the display.

In the second approach, storage is provided as a number of 6-bit memories before the character-generator circuits. After this storage stage, each character is generated and displayed in turn at a rate fast enough to avoid flicker. Each array can only be on for a duty cycle that corresponds to the number of arrays in the display, and must therefore be operated at a relatively high pulse current. The limitation on the number of arrays that can be operated in this fashion thus depends on the maximum pulse current

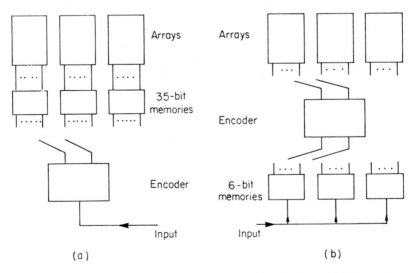

(a)   (b)

FIGURE 6.20. The organization of multicharacter displays: (a) information is stored in 35-bit memories after the encoder and the display operates continuously and (b) information is stored in 6-bit memories before the encoder and the display is scanned

that can be applied to the arrays, as well as the speed at which the circuits can operate.

6.6   Radar and communications

Radar and communications systems can operate over a wide range of the electromagnetic spectrum. Most systems operate on radio or microwave frequencies with wavelengths as short as a few millimetres but there are often advantages to be gained by using higher frequencies such as those in the optical region of the electromagnetic spectrum. Among these advantages are the small transmitter apertures needed for a given beam width and the high modulation frequencies that can, at least in principle, be utilized. There are also a number of disadvantages in using optical radiation. One of the most severe of these is the attenuation of optical radiation in the atmosphere and the relatively low powers available from suitable sources. In general optical systems may be of advantage in situations where a high frequency and a narrow beam width are required, and a range of up to a few kilometres is adequate.

Radar and communications systems can both be considered as consisting of two distinct components: the transmitter and the receiver. Each of these is also conveniently subdivided as shown in Figure 6.21.

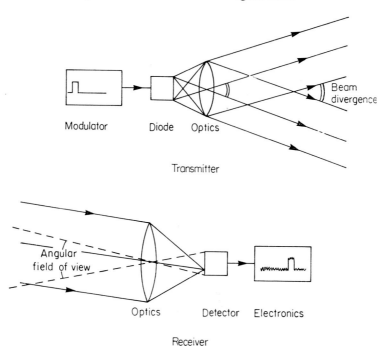

FIGURE 6.21. The basic components of an optical radar or communications system

The transmitter consists of a modulator, which drives the diode or laser, and an optical system to collimate the radiation from the device. The detector comprises an optical system and detector with suitable electronics.

The systems that will be considered will, in most cases, use devices that emit infrared radiation. In some situations, electroluminescent diodes can provide sufficient power for a viable system, but often lasers are required to give increased power and, even here, it may sometimes be necessary to cool the device to obtain adequate performance. At present, gallium arsenide lasers are almost invariably used in laser systems, since they have been developed to a more advanced state than any other semiconductor laser. However, devices with a similar performance but somewhat shorter-wavelength emission can be made from gallium–aluminium arsenide, and these may have advantages in some situations.

6.6.1 Modulators

To obtain high resolution in radar systems or high data rates in a pulsed communications system, a modulator that will fully exploit the potential of a semiconductor diode or laser must be used. This means that the modulator must be capable of producing current pulses with a rise time of about 10 ns or less, and may be required to provide current pulses of up to 100 A amplitude into the low impedance of the device. To achieve these requirements, modulators based on avalanche transistors or silicon-controlled rectifiers (s.c.r.s) have been widely used.

[i] *Avalanche transistors.* The circuit of an avalanche-transistor pulse circuit is shown in Figure 6.22. Using a single transistor, 10 A current pulses can be obtained, but, to obtain higher currents, several devices must be used in parallel.[8] Slight variations in the characteristics of the transistors will

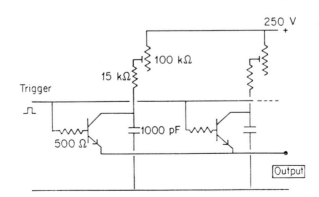

FIGURE 6.22. A pulse-generating circuit using avalanche transistors in parallel

cause them to switch at slightly different times, thus degrading the shape of the current pulse. This can be overcome by adjusting the collector bias voltage of each transistor so that they all switch at the same trigger voltage.

[ii] *Silicon-controlled rectifiers.* Circuits based on the use of s.c.r. switches are capable of producing current pulses of 200 A or more, but the rise times obtained are not as fast as those obtained from avalanche-transistor circuits. A typical circuit is shown in Figure 6.23. The output of this circuit is transformer-coupled so that both the diode and the current-monitoring resistor can be earthed to simplify the electrical design of the transmitter.

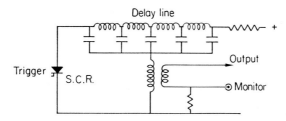

FIGURE 6.23. A pulse-generating circuit using a silicon-controlled rectifier

6.6.2 Cryogenics

The first lasers that were developed could only operate under pulsed conditions and at low temperatures. With the improvements that have been made to devices the need for cooling has decreased, and often adequate performance can be achieved by devices operating at room temperature. However, to obtain the highest possible powers cooling is still an advantage, and useful improvements in performance can be achieved by cooling with thermoelectric coolers or liquid nitrogen.

Liquid-nitrogen temperatures are conveniently obtained by using cooling devices that utilize the Joule–Thomson effect.[9,10] These devices require a supply of nitrogen at a pressure of 200 atmospheres. This high-pressure gas expands through a small orifice and its temperature drops. The cooled gas then exhausts past the incoming gas, which is thus also cooled. Consequently the temperature of the gas emerging from the expansion orifice progressively drops until it liquifies. Devices of this type can produce liquid nitrogen only a few seconds after the gas flow is initiated, and provide a cooling power of a few watts.

6.6.3 Transmitter optics

The design of an optical system for a transmitter presents few problems. The function of this optical system is to collimate the radiation from the source to produce a narrow beam. If the linear dimension of the source is d and a lens of focal length F is used, the divergence of the collimated beam is d/F.

The aperture of the lens must be large to collect as much of the available power as possible, although the size of the lens will often be limited by the maximum permissible size of the system. It is thus instructive to calculate the fraction of available power that is collected by a lens of a given aperture.

Suppose the source has a polar intensity distribution $I(\theta)$. The fraction of available power within a cone of semiangle θ is

$$F(\theta) = \frac{\int_0^\theta I(\theta)\sin\theta d\theta}{\int_0^{\pi/2} I(\theta)\sin\theta d\theta}. \tag{6.1}$$

For a source which emits isotropically, $I(\theta)$ is a constant and

$$F(\theta) = 1 - \cos\theta. \tag{6.2}$$

For a Lambertian source, $I(\theta)$ varies as $\cos\theta$, and

$$F(\theta) = 1 - \cos^2\theta. \tag{6.3}$$

These two functions are plotted in Figure 6.24. In practice, most spontaneous electroluminescent devices will lie between these extremes, and it can be seen, for example, that even an optical system with an $F/1$ aperture will collect only about 15 per cent. of the available power.

In the design of a laser system, the linear shape of the source must be considered. The narrow dimension of the source is 2–5 μm so that a lens of relatively short focal length will give a beam of small divergence. The source

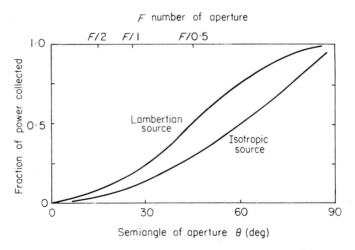

FIGURE 6.24. The fraction of available power collected by an optical system

dimension in the other plane is 10^2–10^3 μm so that a much longer focal length lens must be used to give the same beam divergence.

The radiation polar diagram for a semiconductor laser shows a divergence of about $\pm15°$ in the junction plane and $\pm20°$ in the other plane so that a lens of aperture at least $F/2$ is needed to collimate the radiation.

In a laser system, the optics can be designed to take advantage of the linear shape of the radiation source. Instead of using a single spherical lens, two cylindrical lenses can be used, as in Figure 6.25. The shorter-focal-length lens collimates radiation in the plane perpendicular to the junction, while a

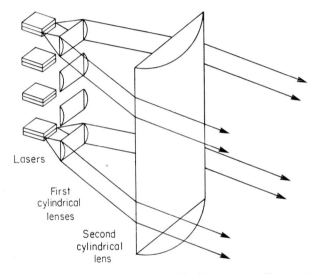

FIGURE 6.25. The use of cylindrical lenses to collimate the
radiation from several lasers

second lens of longer focal length collimates in the junction plane. Since the divergence of radiation in the junction plane is smaller than that in the perpendicular plane, the second lens can have a smaller aperture than the first. Using this system, several lasers can be combined as shown to give a single collimated beam.

6.6.4 Receiver optics

The optical system of a receiver must be such as to collect as much as possible of the desired radiation but reject unwanted background radiation. This implies that it may be necessary to restrict the angle of view of the receiver, although if this angle is restricted too much it may be difficult to locate the source. Background radiation can also be minimized by using a suitable optical filter. The bandwidth of the filter can be approximately 50 Å for use with lasers, although a band width of about 200 Å will be necessary with incoherent sources.

6.6.5   Detectors

The performance of an optical radar or communications system will often be limited by the performance of the detector that is used.[11] In general, the choice of a detector for the near-infrared region of the spectrum will be between some form of photomultiplier or photodiode. This choice may be governed by considerations such as reliability or convenience, but these factors will be ignored in this section, and the choice of a detector to give the optimum performance will be considered.

[i] *Noise performance of detectors.* The detectors that are of interest here can be considered as devices that absorb radiation and produce a photo-current. The sensitivity S of such a detector is then defined as the photo-current produced by unit radiation power at a specified wavelength.

An incident radiation power P will produce a photocurrent I which is given by

$$I = SP. \tag{6.4}$$

This photocurrent will produce a signal power in a load resistor R, and this signal must be detected in the presence of any noise power that is also present in the resistor. Thus the overall requirement of the detector is that the signal-to-noise ratio should be as large as possible so that the smallest possible incident power is detected. The noise power will consist of components due to shot noise on the current flowing through the detector and thermal noise in the load resistor.

[a] Shot noise. In most systems, a detector will receive a background radiation power P_B, which will produce a photocurrent I_B. This current will be superimposed on the dark current I_D, which flows through the detector when no radiation is received. Thus the total detector background current is

$$I_0 = I_D + SP_B. \tag{6.5}$$

The noise power developed by this current in a load resistor R is given by

$$p_S = 2eI_0RB \equiv i_S^2 R \tag{6.6}$$

where B is the bandwidth and i_S is the equivalent noise current.

[b] Thermal noise. The thermal noise power developed in a resistor R is given by

$$p_T = 4kTB. \tag{6.7}$$

[c] Noise equivalent power. The total noise power in the load resistor is given by

$$p_0 = 2eI_0RB + 4kTB. \tag{6.8}$$

It is convenient to consider the incident power P_0 that would produce a signal power equal to this noise power and regard P_0 as a measure of the minimum radiation power that can be detected. Thus, from equations (6.4) and (6.8):

$$P_0 = \frac{1}{S}\sqrt{\frac{p_0}{R}} = \frac{\sqrt{B}}{S}\left(2eI_0 + \frac{4kT}{R}\right)^{\frac{1}{2}} \tag{6.9}$$

and thus P_0 may be defined as the total noise equivalent power of the detector.

If the second term in equations (6.8) and (6.9), i.e. the term relating to thermal noise, can be ignored, P_0 is proportional to the square root of the detector bandwidth B. In this situation, it is usual to define the noise equivalent power (n.e.p.) of a detector as the radiation power that will produce a signal power equal to the detector noise power in a bandwidth of unity. Thus:

$$\text{N.E.P.} = \frac{P_0}{\sqrt{B}}. \tag{6.10}$$

If the detector noise is due to the detector dark current I_D, then, from equation (6.9).

$$\text{N.E.P.} = \frac{1}{S}\sqrt{(2eI_D)}. \tag{6.11}$$

Although the noise equivalent power of a detector is a parameter that is frequently quoted, it is the total noise equivalent power, given by equation (6.9), that determines the performance of a system, since this takes into account not only the noise due to the detector dark current, but also that due to background radiation and the load resistor.

If the detector incorporates some form of current amplification, as in a photomultiplier or avalanche photodiode, both the signal current and the background current are multiplied by a factor m. The shot-noise current is also multiplied by m, and the shot-noise power by m^2. Thus, using equations (6.4) and (6.6), equation (6.9) becomes

$$P_0 = \frac{\sqrt{B}}{mS}\left(2eI_0 m^2 + \frac{4kT}{R}\right)^{\frac{1}{2}} \tag{6.12}$$

$$= \frac{\sqrt{B}}{S}\left(2eI_0 + \frac{4kT}{m^2 R}\right)^{\frac{1}{2}} \tag{6.12a}$$

and it can be seen that the multiplication process has reduced the effect of the thermal resistor noise on P_0.

If the signal and noise currents are multiplied by different factors m and m', equation (6.12) becomes

$$P_0 = \frac{\sqrt{B}}{S}\left(2eI_0\frac{m'^2}{m^2} + \frac{4kT}{m^2 R}\right)^{\frac{1}{2}}. \tag{6.12b}$$

Equations (6.12a) and (6.12b) will be used in a subsequent section to predict and compare the noise performance of several detectors.

From equation (6.12), it is apparent that the load resistance should, if possible, be large enough to make the second term in this equation negligible. If this can be achieved, P_0 increases as $\sqrt{B}$. However, the maximum load resistance that can be used R_m is determined by the required time constant or bandwidth of the detector system B and the capacitance C associated with the detector.

Thus

$$R_m = \frac{1}{CB}. \tag{6.13}$$

If, in these circumstances, the second term in equation (6.9) is still dominant, P_0 increases as B. The conditions in which these two regimes are applicable depends on the detector and the background power that it receives, as well as the bandwidth required by the system. These conditions will be considered in more detail in the sections that consider specific detectors.

[d] Detectivity. The detectivity D^* of a detector is defined as the reciprocal of the noise equivalent power of a detector with a unit sensitive area.

Since the dark current of a detector is, in general, proportional to the area A_D, the equivalent noise current, and hence the noise equivalent power, will be proportional to $\sqrt{A_D}$, and hence

$$D^* = \frac{\sqrt{A_D}}{\text{N.E.P.}} . \tag{6.14}$$

[e] Photon-limited conditions. If the sources of noise that have been described can be reduced to negligible values, for example by cooling the detector to reduce the dark current, the signal-to-noise ratio will be determined only by the shot noise on the signal current I. The signal-to-noise ratio is then given by I^2/i^2, where i is the shot-noise current. Following equations (6.4) and (6.6),

$$\frac{I^2}{i^2} = \frac{S^2 P^2}{2eBSP} = \frac{SP}{2eB} \tag{6.15}$$

and the performance of the detector is limited by its sensitivity and bandwidth. This represents the ultimate performance of the detector, and is known as the photon or quantum limit. The implication of equation (6.15), in physical terms, is that a signal can only be detected if at least one photoelectron is collected in each half cycle of the bandwidth B.

[ii] *Photomultipliers.* Several photomultipliers with a useful response extending into the near infrared are available. The spectral response of some of the photocathodes commonly used is shown in Figure 6.26. The S1 cathode has the greatest sensitivity at 9000 Å, but even this has a low

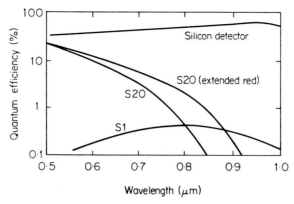

FIGURE 6.26. The spectral response of some infrared photomultiplier cathodes and a silicon photodiode

quantum efficiency of about 3×10^{-3} electrons per photon at this wavelength. Specially selected S20 photocathodes with an extended red response have a comparable performance, but this is strongly dependent on wavelength and the individual devices. The low quantum efficiency of the photocathodes are obviated by the low dark current of the devices. This would be, typically, 10^{-12} A, and could be reduced by a factor of 10^3 by cooling to 200 K.

The dynode chain in a photomultiplier provides current gain without appreciably degrading the noise performance of the device, and this amplified current develops a signal across a load resistor. In general, the gain is high enough for thermal noise in the load resistor to be neglected, and the noise performance of the photomultiplier can be predicted from a knowledge of its sensitivity and dark current.

The total noise equivalent power as a function of bandwidth for a photomultiplier with a cathode sensitivity of 3×10^{-3} A/W and a dark current of 10^{-12} A is shown in Figure 6.27. These device parameters correspond to a typical S1 photomultiplier, and the lower curve applies when no background radiation reaches the detector. If the detector receives background radiation, the noise due to this radiation can be determined from the known sensitivity of the photomultiplier. Curves for representative values of background power are also shown in Figure 6.27.

[iii] *Silicon photodiodes.* In a semiconductor photodiode, detected radiation generates a current that is in the conventional reverse direction of the diode. The current–voltage characteristics of an illuminated device is thus as shown in Figure 6.28, and the spectral response is shown in Figure 6.26.

If the device is operated with no external bias and the photocurrent fed into a load resistor, the load line will be as shown in Figure 6.28 (line A), and power can be extracted from the diode. This corresponds to the operation

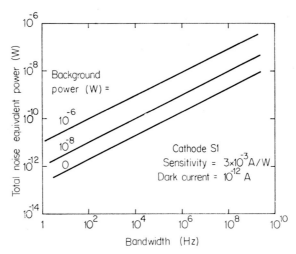

FIGURE 6.27. The noise performance of a photo-multiplier

of a photovoltaic cell or solar cell such as is used to convert solar energy to electrical energy.

For use as a fast and low-noise detector, a *p–n* junction is operated under reverse bias (load line B in Figure 6.28). Devices must be designed with this operation in mind, and must also have a low capacity and leakage current.

The silicon *p–i–n* photodiode has been developed to meet these requirements. The *p–i–n* structure gives a wide junction depletion region, in which photons are absorbed, so that the device has a high quantum efficiency.

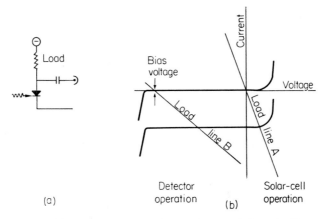

FIGURE 6.28. The current–voltage characteristics of silicon photodiodes

A typical detector has a quantum efficiency as high as 60 per cent. at 9000 Å, a dark current of about 10^{-8} A and a capacitance of 10 pF (for a 2 mm-diameter device).

The noise performance of a *p–i–n* detector can be calculated as in Section 6.6.5, and is shown, for the device specified above, in Figure 6.29. At small bandwidths (less than 3 kHz for this example), the total noise equivalent power is due to shot noise on the diode leakage current. At bandwidths greater than 3 kHz, thermal noise in the load resistor is more important.

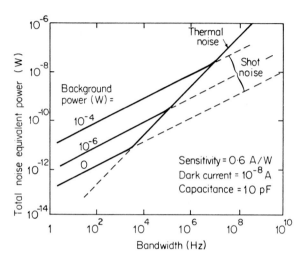

FIGURE 6.29. The noise performance of a silicon *p–i–n* photodiode

The effect of background radiation is to increase the shot noise in the detector and to increase the bandwidth at which thermal noise becomes dominant.

[iv] *Silicon avalanche photodiodes.*[12] The avalanche multiplication process, which causes breakdown when a diode is reverse biased, can be made to multiply the photogenerated carriers in a photodiode. The diode must be carefully constructed to give uniform breakdown over the whole junction area, and is operated under a reverse bias close to this breakdown point. With a well designed device, useful multiplication factors m of about 100 can be achieved.

The noise current in the detector is also amplified, and it can be shown that this is multiplied by a factor $m' = m^x$ where x depends on the relative efficiency with which electrons and holes cause impact ionization; $x \approx 1.15$ for a silicon device, and higher than this for most other materials. This excess noise multiplication arises because electrons *and* holes take part in the multiplication of a steady-state signal, including a background-illumination signal. With a short, transient signal, only electrons are collected by the junction, and they alone take part in the multiplication process. Thus steady-state noise is amplified more than a transient signal.

From equation (6.12), it can be seen that the multiplication process in an avalanche diode reduces the effect of thermal noise in the load resistor. However, since the shot noise in the detector is amplified by a greater factor than the signal, there is a limit to the useful multiplication that can be used.

The noise performance of a typical avalanche photodiode is shown in Figure 6.30. The device has a quantum efficiency of 60 per cent. at 0·9 μm, a capacitance of 10 pF, and operates at a signal-multiplication factor of 100. The bulk leakage current in the junction region of such a device may be as low as 5×10^{-11} A, but is multiplied by a factor $m' = 100^{1·15} = 200$ to give a measured dark current of 10^{-8} A. Thus the total noise equivalent power is given [from equation (6.12)] by

$$P_0 = \frac{\sqrt{B}}{0·6} \left[2e \frac{200^2}{100^2} (5 \times 10^{-11} + 0·6P) + \frac{4kT}{100^2 R} \right]^{\frac{1}{2}} \text{watt.} \qquad (6.16)$$

It can be seen from Figures 6.29 and 6.30 that the multiplication process has reduced the effect of resistor noise on P_0 by a factor of 100, but has increased the effect of shot noise due to background radiation by a factor of two. Thus the use of the avalanche multiplication process is advantageous

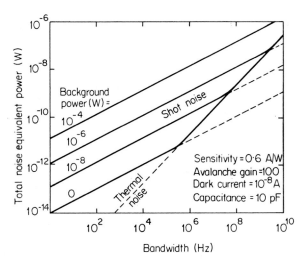

FIGURE 6.30. The noise performance of a silicon avalanche photodiode

only in conditions involving a large bandwidth and low background radiation.

[v] *Comparison of detectors.* The principles that have been discussed in the previous sections can be used to compare the noise performance of detectors as a function of the system parameters, bandwidth and background-radiation power. The general form of such a comparison is shown as the locus of system parameters for which photomultipliers, silicon photodiodes and avalanche photodiodes have equivalent performance (Figure 6.31). The exact details of such a plot will depend on the properties of the device that are compared. Taking the detectors that have been considered as

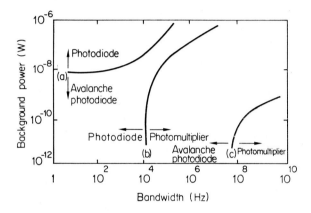

FIGURE 6.31. A schematic comparison of the noise performances of an S1 photomultiplier, a silicon *p–i–n* photodiode and a silicon avalanche photodiode; the curves show the locus of operating conditions for which (a) *p–i–n* photodiode and avalanche photodiode, (b) *p–i–n* photodiode and photomultiplier and (c) avalanche photodiode and photomultiplier have equivalent performance; these curves only show the general form of such a comparison

examples, with no background radiation the performance of a photomultiplier will be superior to that of a *p–i–n* diode at frequencies greater than 10^8 Hz. In the presence of background radiation, these frequencies increase qualitatively as shown, and the region in which a photomultiplier is ultimately superior is confined to the high-bandwidth, low-background-power region of the figure.

In practical systems, solid-state devices have further advantages over photomultiplier tubes. These advantages include ruggedness and lower-voltage operation. Thus, in the majority of situations, a solid-state device will be preferred and will give superior performance compared with a photomultiplier at near-infrared wavelengths.

6.6.6 *Optical radar and range-finding systems*

In a radar or range-finding system, a transmitter illuminates a target with a pulse of radiation. This radiation is scattered by the target, and the scattered radiation is detected by a low-noise detector. A system will, in general, utilize readily available radiation sources and detectors, and must be designed to give an optimum performance subject to the limitations of these components and any particular requirements imposed on the system.

Although a comprehensive discussion of the principles of radar is beyond the scope of this book, some aspects that are particularly relevant to optical radar will be considered, and these will be illustrated with a brief description of some practical systems. A more detailed discussion will be found in Reference 13.

[i] *Beam widths.* In an optical radar system, the transmitter produces a beam with a certain divergence, and the target may or may not intercept all the radiation. The detector will also have a finite field of view, which may be larger or smaller than the illuminated target areas. Thus there are four situations which must be considered. These are illustrated in Figure 6.32 and summarized in Table 6.2. For this discussion, it is assumed that the transmitter and detector are both in a direction normal to a target at range R, and that the target scatters according to Lambert's law.

Suppose the transmitted power W is emitted into a beam of solid angle

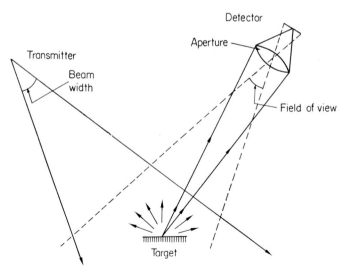

FIGURE 6.32. The basic geometry of a radar system

TABLE 6.2. The radar range equations.

Target Area		Illuminated Area	Scattered Power	Detected power	
				Small field of view	Large field of view
A	$>$	$\Omega_T R^2$	W	$W \dfrac{\Omega_D R^2}{\Omega_T R^2} \dfrac{A_D}{\pi R^2}$ $\sim \dfrac{\Omega_D}{\Omega_T} \dfrac{A_D}{R^2}$	$W \dfrac{A_D}{\pi R^2}$ $\sim \dfrac{A_D}{R^2}$
A	$<$	$\Omega_T R^2$	$W \dfrac{A}{\Omega_T R^2}$	$W \dfrac{A}{\Omega_T R^2} \dfrac{\Omega_D R^2}{A} \dfrac{A_D}{\pi R^2}$ $\sim \dfrac{\Omega_D}{\Omega_T} \dfrac{A_D}{R^2}$	$W \dfrac{A}{\Omega_T R^2} \dfrac{A_D}{\pi R^2}$ $\sim \dfrac{A A_D}{\Omega_T R^4}$

Ω_T. At the target, the cross-sectional area of this beam is $\Omega_T R^2$, so that the power that is intercepted and scattered by a target of area A is

$$W \qquad \text{if } A > \Omega_T R^2 \qquad (6.17)$$

or

$$\frac{WA}{\Omega_T R^2} \qquad \text{if } A < \Omega_T R^2. \qquad (6.18)$$

These situations correspond to the target filling or not filling the transmitted beam. Some of the radiation scattered by the target will be collected by the detector whose field of view is the solid angle Ω_D and aperture is A_D. The power collected by the detector can be calculated as the product of three factors:

$P = W \times$ (fraction of power back-scattered from target)

$\qquad \times$ (fraction of illuminated target area within detector field of view)

$\qquad \times$ (fraction of radiation collected by detector of area A_D at range R).

The result of these calculations, for the four situations outlined above, are given in Table 6.2. It can be seen that the range dependence of the received power reduces to two alternatives:

If the target is small compared with both the transmitter beam width and the detector field of view, as is usually the case,

$$P \sim \frac{1}{R^4}. \qquad (6.19)$$

In other situations

$$P \sim \frac{1}{R^2}. \qquad (6.20)$$

G

[ii] *Atmospheric attenuation.* The range equations given in Table 6.2 must be modified to take account of the attenuation of the transmitted power by the atmosphere by a factor $\exp(-2\alpha R)$, where α is the appropriate absorption coefficient. Since most of the systems that are of interest operate at a wavelength of less than 1 μm, the attenuation of radiation is not very different from that of visible radiation, and the systems have a poor performance in fog, cloud or rain. For the best conditions, with a visibility of 40 km, $\alpha = 0.05$ km^{-1}.

[iii] *Background power.* The detector will receive radiation due to any ambient illumination falling on the target. If the field of view of the detector is larger than the target, it may also receive radiation from the background surrounding the target. The total background radiation will contribute to the shot noise in the detector, and must therefore be kept to a minimum. Besides choosing the detector in accordance with the principles discussed in Section 6.6.5, the background noise level can be reduced in two ways—an optical filter can be used to reject background radiation, and the detector field of view can be restricted. In the latter case, if the detector field of view is too restricted, target acquisition can become difficult.

[iv] *Bandwidth and source power.* To determine the range of a target the time that elapses between the transmission of a pulse and its detection must be measured. It will be assumed that such measurements are to be made at a predetermined frequency, and that a single pulse of radiation is to be detected. In a sophisticated system, this may be a serious oversimplification, but it will serve to illustrate the principles which are involved.

To determine range to an accuracy Δx, the time of flight of the radiation pulse must be determined to an accuracy $\Delta x/c$, where c is the velocity of light. To achieve this, the radiation must have a pulse length τ of this order, and the detector must have a bandwidth $B \sim c/\Delta x$. For example, a resolution of 10 cm requires a pulse length of 1 ns and a detector bandwidth of 1 GHz.

Although the detector bandwidth must be sufficient to give the required range resolution, the optimum bandwidth is determined by the characteristics of the radiation source and the dominant noise mechanism in the detector (Figure 6.33).

Suppose W is the source power and P_0 the detector noise equivalent power. P_0 has the form [from equations (6.12) and (6.13)]

$$P_0 = (\alpha B + \beta B^2)^{\frac{1}{2}} \tag{6.21}$$

where α, β are constants. The signal-to-noise ratio of the system will be determined by the ratio

$$\frac{W}{\sqrt{(\alpha B + \beta B^2)}}.$$

If the peak power of the source is limited, the best performance will be

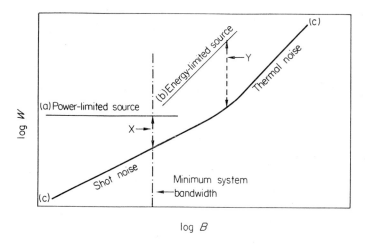

FIGURE 6.33. Conditions for the optimum bandwidth of an optical radar system. Curves (a) and (b) show the received signal power as a function of bandwidth for (a) power-limited and (b) energy-limited sources. Curve (c) shows the noise power as a function of bandwidth. Optimum performance occurs at the bandwidth that gives the maximum signal-to-noise ratio that is compatible with the required frequency response

obtained when the bandwidth is kept to the minimum value required to give adequate range resolution (X in Figure 6.33).

Alternatively, in some situations the pulse energy available may be limited so that a higher power can be obtained with a shorter pulse length. The detected power will increase as $1/\tau$, or B, and the signal-to-noise ratio is determined by

$$\frac{B}{\sqrt{(\alpha B + \beta B^2)}}.$$

In these circumstances, the best performance is obtained by increasing the bandwidth and peak power of the system to the point at which thermal noise in the load resistor dominates the detector shot noise (Figure 6.33: Y).

[v] *Experimental systems.* A number of experimental optical radar and range-finding systems have been developed. Most of these have used gallium arsenide lasers as the source of radiation, and two systems will be described briefly.

Birbeck and Hambleton[14] have described a system that was designed as a low-level aircraft altimeter. This used a gallium arsenide laser at room temperature and a silicon *p–i–n* photodetector. The system was operated with these components at altitudes of up to 500 m, but, with more recently developed lasers and an avalanche photodetector, this range could be increased by an order of magnitude.

A system described by Hansen[15] has been used as a high-resolution rangefinder. A gallium arsenide laser was operated to give an output-pulse power of 2 W with a rise time of 1 ns. A photomultiplier was used as the detector, and a range resolution of 7·5 cm was obtained. When used in daylight conditions, with a 100 Å filter, the detection system was limited by noise due to background radiation, and ranges of up to 90 m could be measured.

6.6.7   *Communications systems*

Many of the basic principles of optical radar systems apply to communications systems, but there are also additional features which must be considered.

[i] *Modulation*. The high frequency of optical radiation ($\sim 10^{14}$ Hz) implies that extremely high bandwidths should be possible in optical communications systems. This potential has by no means been exploited because the available modulation and detection techniques are limited to much lower frequencies.

One of the advantages of electroluminescent devices is that their output can be modulated by modulating the current which flows through the device. Both amplitude- and pulse-modulation techniques can be used in this way.

Amplitude modulation can be applied to incoherent diodes at frequencies up to about 100 MHz. The maximum frequency is limited by the rise and decay time of spontaneous recombination in the device. Under carefully controlled conditions semiconductor lasers have been amplitude modulated at frequencies exceeding 10 GHz, but more practical systems have been limited to frequencies of about 10 MHz.

Various forms of pulse modulation have been used with both electroluminescent diodes and lasers. In the latter case, pulse modulation is usually more viable than amplitude modulation, since the c.w. operation required for amplitude modulation can only be obtained under very stringent conditions. Lasers can be pulsed with a rise time of a few nanoseconds. Pulse repetition frequencies approaching 10 MHz can be obtained if the laser operates at a 10 per cent. duty cycle and 100 MHz if the laser will approach c.w. operation. To obtain the faster rise times required for higher pulse frequencies, it is probably necessary to bias the laser with a d.c. drive to just below threshold and superimpose the modulating current. In this way pulse repetition rates of 1 GHz are feasible.

[ii] *Transmission*. The transmitted signal may be radiated freely through the atmosphere or transmitted along some form of fibre-optic path.[16]

The problems that arise using atmospheric transmission are similar to those encountered in radar systems in that the performance is dependent on the atmospheric conditions. However, the transmitter beam width and detector field of view can be kept small, particularly if the system is stationary, to obtain the best possible signal-to-noise ratios.

Two alternative forms of fibre-optic communication are possible (Figure 6.34). In the first of these, radiation is transmitted through a glass fibre whose diameter is large compared with the radiation wavelength. The core of the fibre is clad with a glass of lower refractive index so that total internal reflection confines the radiation to the fibre. In general, a large number of possible paths exist for the radiation, and the transmission time for each of these paths will be different. This will set a limit to the range and frequencies that can be used before successive pulses become intermixed and indecipherable.

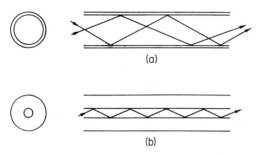

(a)

(b)

Figure 6.34. Fibre-optic communication channels: (a) multimode and (b) single mode

For a high-frequency optical communication system, a single-mode optical waveguide can be used. This consists of a thin glass core surrounded by a cladding whose refractive index is about 1 per cent. lower than that of the core. Such a waveguide can be designed to propagate a single mode. The frequency limit of this form of transmission is now set by the dispersion of the glass and the spectral linewidth of the radiation source. For example, a refractive index variation of 1 in 10^4 might occur over a wavelength interval of 100 Å, and this would give a transit time spread of 1 ns over a range of 1 km. This would limit the usable bandwidth to less than 1 GHz.

It is also necessary to keep the attenuation of radiation in the optical waveguide to a low value. This entails the preparation of glass of high purity, and the subsequent manufacture of high-purity fibres. The best grades of commercial glass give an attenuation of about 150 dB/km, but careful purification to keep the concentration of certain impurities (e.g. iron) to less than 1 part in 10^6 improves this figure, and, at present, fibres with an attenuation of only 15 dB/km can be made.

[iii] *Radiation sources.* A source that will operate at a high duty cycle, preferably continuously, is needed to obtain a high communications bandwidth. At present, semiconductor lasers have only achieved high duty cycles in the laboratory, and practical systems can only utilize devices with a 10 per cent. duty cycle and a pulse-repetition frequency of about 10 MHz.

Incoherent sources operate continuously, and can be modulated at frequencies approaching 100 MHz. However, the power available from these sources is modest, and their peak radiance is much lower than that of a laser. Thus the choice between a laser and an incoherent diode is not clear cut, and each application must be assessed in detail.

6.7 Conclusions

Semiconductor lasers have been available for nearly ten years, but, at present, have not found widespread viable applications. However, devices can now be made to operate at a high duty cycle at room temperature, and this may well increase their use, particularly in the field of communications.

Over the same ten years, electroluminescent diodes emitting visible radiation have demonstrated their potential in display systems and have become widely accepted. Their usefulness would be extended considerably if colours other than red were more widely available. Considerable research effort is aimed in this direction, and it seems that at least red- and green-emitting devices will soon be incorporated in display systems.

6.8 References

1. R. L. Hartman, B. Swartz and M. Kuhn, *Appl. Phys. Lett.*, **18**, 304 (1971).
2. K. L. Konnerth, J. C. Marinace and J. C. Topalian, *J. Appl. Phys.*, **41**, 2060 (1970).
3. R. L. Longini, *Solid State Electron.*, **5**, 127 (1962).
4. R. D. Gold and L. R. Weisberg, *Solid State Electron.*, **7**, 811 (1964).
5. D. P. Cooper, C. H. Gooch and R. J. Sherwell, *J. Quantum Electron.*, **2**, 329 (1966).
6. M. Ettenberg, H. S. Somers, H. Kressel and H. F. Lockwood, *Appl. Phys. Lett.*, **18**, 571 (1971).
7. W. Rosenzweig, *IEEE. J. Solid-State Circuits*, **5**, 235 (1970).
8. J. P. Hansen and W. A. Schmidt, *Proc. IEEE.*, **55**, 216 (1967).
9. 'Minicooler' data sheet, Hymatic Engineering Co. Ltd., Redditch, Worcs., England.
10. V. M. Farmer and D. P. Forse, *Infrared Physics*, **8**, 37 (1968).
11. H. Melchior and M. B. Fisher, *Proc. IEEE.*, **58**, 1466 (1970).
12. K. M. Johnson, *IEEE. Trans. Electron Dev.*, **13**, 164 (1966).
13. B. S. Goldstein and G. F. Dalrymple, *Proc. IEEE.*, **55**, 181 (1967).
14. F. E. Birbeck and K. G. Hambleton, *J. Sci. Inst.*, **42**, 541 (1965).
15. J. P. Hansen, *Proc. IEEE.*, **57**, 854 (1969).
16. Special issue of *Proc. IEEE.*, **58**, Oct. 1970.

Author Index

189

Subject Index

193